THE CREATION OF QUABBIN RESERVOIR

The Death of the Swift River Valley

by

J. R. GREENE

Copyright © 1981 By J.R. Greene

This volume or parts thereof may not be reproduced without written permission of the author.

Printed By:

THE TRANSCRIPT PRESS
32 Freedom Street
Athol, Mass. 01331

Second Edition, First Printing 1987
ISBN 0-9609404-0-5

Front Cover Illustration: Enfield Congregational Church

Back Cover Illustration: Quabbin Reservoir

THE CREATION OF QUABBIN RESERVOIR
CONTENTS

Preface

Part I — The Valley

    1. Origins   1
    2. Growth   3
    3. Decline   5

Part II — Boston's Historic Need for Water

    1. Early Developments   9
    2. The Board of Health and its Investigations   12
    3. Wachusett Reservoir   15
    4. More Cities; More Water   18
    5. The Joint Board Report   22

Part III — The Water Fight

    1. Politics   25
    2. The General Court Tries Its Hand   27
    3. Another Investigation   31
    4. A Third Investigation   38
    5. More Politics   46
    6. The Swift River is Taken   52

Part IV — Construction

    1. The Work Begins   58
    2. To the Supreme Court   62
    3. Exodus   66
    4. The Dams   73
    5. Four Towns Cease to Exist   79
    6. The River Rises   85

Part V — Cup Bearer to All Mankind

    1. Effects of the Reservoir   89
    2. More Users - More Diversions?   94

Part VI — Appendices

    1. Footnotes   99
    2. Bibliographic Essay   101
    3. Population Tables   106
    4. Index of Persons   107

Maps and illustrations are located between pages 40 and 41.

THE CREATION OF QUABBIN RESERVOIR

The Death of the Swift River Valley

Time is busy in levelling hills, abrading garden terraces, flattening grave mounds, and abasing old families.

> Francis Underwood in *Quabbin: The Story of a New England Town*

But say not death has come to me
Beauty, not dust, you'll find
For I stand among my towering hills
Cup-bearer to all mankind.

> From a poem quoted on the book jacket of *Quabbin: The Lost Valley*

(Once) we more or less told the citizens what we were going to do, then went ahead and did it. That world has come to an end.

> Thomas N. Urban,
> Former Mayor of Des Moines, Iowa

DEDICATION:

To My Mother

PREFACE

This work is intended as a study of the events that led to the creation of the Quabbin Reservoir in Central Massachusetts. This body of water is unusual for the New England region in its scope. Four towns and several other villages were wiped off the map to create the 38 square mile Quabbin.

No complete history of the towns affected by the reservoir will be attempted here: detailed sketches of the towns may be found in other works. Certain aspects of the area's history which have been lightly touched upon before will be examined as they point to why the four towns suffered the fate they did.

Although engineers are considered in depth, as are some technical aspects of reservoir construction, this is not a technical study. Such material filled out many an article in engineering journals before the Second World War.

The stories of individuals are of concern here only as they affect the story of the reservoir. This work is meant for those concerned about the process of the events involved, not as a rehash of local gossip or genealogy.

There is a moral question here, or rather, several, and conclusions will be drawn accordingly. After all, is any historical work truly objective? The fact that the state can take away a person's home, land, and physical environment can spawn numerous philosophical arguments.

This was not written as a formal academic study of the subject, but footnotes and other "scholarly" apparatus are used to present a properly documented work. The use of certain "historical approaches" was rejected as being irrelevant to the topic, while it is too complex to be a simple chatty local history. With these points in mind, it is hoped that the reader can view this work from an angle free of misconceptions.

The writer spent countless hours over the last six years travelling, researching, and interviewing to gather information in compiling this work. I wish to thank all those who have assisted me in this task, especially those who agreed to be interviewed (see Bibliographical Essay).

I also wish to thank the following institutions: Athol Public Library, especially Mrs. Alice Newton; Barre Public Library; Forbes Library in Northampton, especially Mr. Stanley Greenberg; Ingalls Library in Boxford; Jones Library in Amherst, especially Mrs. Winifred Sayer; the Massachusetts State Library and Archives; Springfield Public Library; University of Mass., Amherst Campus Tower Library, especially Mr. John Gnatek; Ware Public Library; Wheeler Library in Orange; Widener Library of Harvard University; and the Worcester Public Library.

Also the Swift River Valley Historical Society, especially Mrs. Althea Daniels and Mr. Warren Doubleday; the Prescott Historical Society, especially Miss Ellen Coolidge; the Metropolitan District Commission, especially Mr. Harold Willey, Capt. Albert Swanson, Mr. John Copithorne, Mr. Donald

Slongwhite, Mr. Gerard Albertine, Mr. Bruce Spencer, Mr. Roger Lonergan and Mr. Robert Portier.

I also would like to thank some of the individuals who have assisted me in some way, including Mr. Richard J. Chaisson, former Rep. H. Thomas Colo, Mr. Robert Evans, Mrs. Ruth Rice, Ms. Jennifer Spivack, and Sen. Robert D. Wetmore.

The content of this work is the sole responsibility of the author, and not any of the above-mentioned organizations and individuals.

> Athol, Mass.
> July 15, 1981

PART I   THE VALLEY

Chapter One   Origins

Time began for the Swift River Valley as it did for much of New England. 10,000 years ago, glaciers retreated from the region, leaving their peculiar scars upon the land. In the southwestern part of the Central New England Upland, two parallel valleys were cut, separated by the Prescott Ridge. The soil in these valleys was somewhat sandy, reflecting its glacial origin, but the area was well watered.

Three branches of what the white man called the Swift River drained the valleys. Originally, all three streams flowed southward on independent courses. The Middle Branch pirated the East Branch, then the West Branch did the same to the combined stream. The glaciers left many conical hills and mounts in the eastern lowland, affording good views of the surrounding territory. [1]

A branch of the Nipmuc Indians lived in the valley when the first white men came. One of the sachems (chiefs) was reputedly named NineQuabin or NaniQuaben. The prefix Nine or Nani means chief; the Quabin part refers to a well watered place. According to legends, other places in the region, such as Lake Neeseponsett and Pottapaug Pond, received Indian names. Besides these words, there was little left of the natives in the valley by the time of the Revolutionary War. [2]

The first white settlers to obtain land grants in the valley arrived in the 1730s. These people were descended from colonists who had served the Bay Colony in King Philip's War of 1675-6, specifically in defense of Fort Narragansett in 1675. In 1736, the Great and General Court legislature of Massachusetts granted six townships of land, including 18,709 acres of land "lying at a place called Quabbin" in Hampshire County, which became Narrangansett Township No. 4. The land stretched from the junction of the East and Middle Branches of the Swift River in the South to Lake Neeseponsett and Pottapaug Pond in the northeast. [3]

Many of the early settlers of this tract were of Scottish descent. They remembered their place of origin when the town was formally named upon incorporation in 1754. The Duke of Greenwich was honored by the new name of the Quabbin district. Only a few score people lived in the town, as was the case with nearby New Salem and Petersham, both one year old towns. Pelham, to the northwest, was already 11 years old. Between these four townships, the future towns to occupy the Swift River Valley would be spawned.

In the late 18th and early 19th centuries, these towns were primarily farming communities. As the land dominated their physical life, the state Congregational Church was the chief force in their settler's spiritual existence.

Each village center had its green for grazing livestock, and its white church for Sunday worship and town meetings. What few events took place in town were "naturally grouped under the reigns of it's successive ministers." [4]

Greenwich was not so isolated that it ignored the patriotic call to arms in 1775. Six of the town's 800 citizens joined the Continental Army after the battles at Lexington and Concord, and others joined later.

In the Revolutionary War period, several small villages began to form in outer sections of the valley towns. A convenient mill site or good farmland drew the settlers who formed the nucleus of a village. Unfortunately, several miles often separated these people from the church they were bound to attend. Both Greenwich and Pelham had villages which petitioned for the right to form their own parish in the 1780s. A parish had the right to support its own church, but politically it was still part of the parent township.

In 1786, the part of Pelham east of the West Branch of the Swift River formed the East Parish of Pelham. The center was situated at the middle of the ridge running north to south through the parish. A year later, the southwest part of Greenwich became the South Parish of that town. What later became the village of Enfield was located on the Swift River between the southern end of the Prescott Ridge and Great Quabbin Mountain. A small mill and smithy were already located there. [5]

The 1780s saw the Massachusetts farmer suffer from the economic depression that hit after the conclusion of the Revolutionary War. If the state would issue paper money as legal tender, the farmer felt he could pay off his debts and return to prosperity. The legislature did not act on the behalf of the farmers in the western part of the state, so they took matters into their own hands. Courts that attempted to foreclose mortgages were shut down by mobs of angry farmers.

State Senator Samuel Adams, once styled a radical, addressed the western towns in a tone sympathetic but scolding. Greenwich boldly replied that Boston should not object to the means of dissent it had used "when our grievances were less real and more ideal than they are now." [6] Captain Hinds of Greenwich led a company of his townsmen and Pelham men to stop the courts at Northampton. [7]

Later, when a large force was mustered to capture the arsenal in Springfield, Captain Daniel Shays of Pelham East Parish became the leader of the insurgents. Thus a valley man gave his name to what was styled a rebellion. After an unsuccessful attempt to take the Springfield arsenal in early 1787, his force fled across the valley, only to be dispersed in Petersham. This might have been the first incident of wide disagreement between eastern and western parts of Massachusetts, but it wasn't the last.

Chapter Two   Growth of the Valley

In 1800, Greenwich had almost 1,500 residents, eleven school districts, a number of small mills, and two turnpikes traversing it. Residents of the northeastern section of town had joined with citizens of the southern part of Petersham and northwest Hardwick to petition the General Court for a town of their own. On February 18, 1801, their wish was granted, and the town of Dana was incorporated.

Possibly named after Judge Francis Dana, the town was about five by four miles in size, with irregular boundaries. There were two villages in the new town. The beginnings of North Dana sat at the southern end of a long lake feeding the Middle Branch of the Swift River. Dana Common was on a flat stretch of land north of Pottapaug Pond.

Fifteen years later, on February 15, 1816, the old South Parish of Greenwich broke away from the mother town to become the town of Enfield. Land was annexed from Belchertown on the west, so that the West and East Branch valleys joined within the town. The name Enfield was derived from Robert Field, one of the town's leading citizens. The East Branch of the Swift provided excellent waterpower, which was extensively utilized shortly after the creation of the town.

Residents of the southern part of New Salem, and those in the East Parish of Pelham agitated for years to separate from their parent communities. Their desires were granted on January 28, 1822, when the town of Prescott was formed. Legend has it that the inhabitants wanted to name the town Warren, but they were thwarted by the town now bearing that name. [8] This seems unlikely, since the current Warren was not so named until 1834. In any case the name Prescott was derived either from an opponent of Shays' Rebellion or a Revolutionary War hero.

This town, being shaped like an inverted L, consisted of about 12,700 acres. Three villages of note were located in Prescott. Prescott Hill, southern end of town. At the northern edge of town sat North Prescott, which was partly within New Salem. About halfway between these two was Atkinson Hollow, the location of the noted Atkinson Tavern. Most of the town sat upon a ridge running north to south, with the West Branch of the Swift forming the west boundary. The Middle Branch of the Swift ran through the northeast edge of town.

Despite the losses of terrirory and people, Greenwich prospered. By the 1850s, the town housed a scythe factory, two saw mills, one grist mill, a plating factory, and firms producing farm tools, carriages, buttons and matches. [9] Most of this activity was centered in Greenwich Village, northeast of the center on the East Branch of the Swift. Both villages had post offices.

Enfield took advantage of its position on the lower Swift River shortly after becoming a town. The main export of the town had been whetstones, derived from soapstone deposits on Quabbin Mountain. Factories producing wool, cards, boxes, and pegs were set up in the main village. A short distance upstream, another village sprang up around factories making cotton and wool products. This eventually became known as the Upper Village, or Smith's after the mill-owning family. Stores occupied both villages, while Enfield had a post office. [10]

Dana was able to take advantage of its natural resources to produce potash and quarry granite and soapstone. Hats, piano legs, and pocket books were produced in factories in North Dana. The town added parts of Hardwick and Petersham in 1842, gaining the village of Storrsville. Four religious societies shared one building in the center of town. Hosea Ballou, the noted pioneer of Universalism, spent a decade preaching in Dana in the early 1800s. By the early 1850s, both Dana and North Dana had a store and post office.

Prescott, having the most rugged topography of the four towns, never developed much industry. Besides agricultural products (the foremost being cheese), palm leaf hats were made in many homes, and charcoal was produced in a kiln near the Middle Branch of the Swift. A bobbin factory operated just across the West Branch in Pelham, undoubtedly employing Prescott residents. [12] By the 1850s both Prescott and North Prescott had post offices, stores, and churches, but the population already was beginning to decline. This was a portent of things to come for the other towns.

A nearby section of the state should be noted here, as it would become affected by the same events that would overwhelm the Swift River Valley. This area was the upper Ware River Valley, in the towns of Barre, Oakham and Rutland.

These three towns were originally all part of Rutland, whose land was purchased from Indians in 1686. Rutland grew rapidly, and was incorporated in 1722. In 1762, Oakham, the "west wing" of Rutland, became a separate town, and Barre (first named Hutchinson) incorporated in 1774.

The Ware River flowed through eastern Barre for quite a distance, barely touching Oakham. Many streams in Oakham and Rutland were part of the Ware watershed, and these provided numerous mill sites. The famed Barre Falls in East Barre was the site of many mills. This activity flourished there until the late 1800s, when the south parts of Barre became the dominant manufacturing areas in town.

A turnpike (the sixth Massachusetts) and later railroads helped develop a village in the northwest part of Oakham. This was Coldbrook Springs. It housed a mill, church, school, and post office-store combination.

Rutland continued to prosper even after losing lands to Barre and Oakham. A textile mill on the East Branch of the Ware was the central feature of North Rutland. A large woolen mill was the dominant feature of West Rutland, which was adjacent to a number of ponds. Agriculture was still successfully carried out in all three towns, especially Rutland.

As it turned out, the decade before the Civil War was to be the high point in the existence of the Swift Valley towns. As Francis Underwood observed in 1893

> While all these mills and shops were flourishing, the stores were prosperous, new dwellings were built, and new faces appeared on Sunday in the meeting house, Quabbin appeared to have what in modern times is called a "boom", but it was not to endure. [13]

Chapter Three   Decline

As the industrial revolution had spawned the industries of the valley towns, so too did it spawn the instruments of the valley's decline. The coming of the railroad symbolized the continuing effects of advancing industry on the isolated valley. Because of its north-south orientation, the valley was bypassed when the first railways were laid across Massachusetts. In 1841, the first rail link between Boston and Albany passed 12 miles south of Enfield. By the end of the decade, another east-west line passed through Athol and Orange a few miles north of the valley. These lines were close enough to encourage the slow exodus of valley residents which began at this time, [14] but too far away to encourage additional industry or settlement.

This judgement would seem to be reinforced by the effects of the construction of a railroad through parts of all four towns in the early 1870s. The new railway was received with enthusiasm, for many local residents invested in the line's stock. [15] The line connected the towns with Athol on the north and Springfield to the south, both of the latter being located on trunk lines.

The railroad did not live up to the expectations of the investors. It ran with "indifferent success" even after it had become a branch line of the Boston & Albany rail system. [16] The valley remained relatively isolated, as evidenced by the "decrease in the population and the number employed in manufacturing there."[17] The local railroad had done little to reinforce existing business, and attracted no new industries, except for ice harvesting on local ponds.

The gradual development of a national network of railroads during the last decade of the 1800s made larger markets practicable for different industries. This encouraged the consolidation of businesses into large combines. Such centralization resulted in a marked savings by achieving better economies of scale. [18] These savings, and the perfection of more efficient modes of power generation made the small, water driven mills of the Swift River towns obsolete or noncompetitive. Some of the small industries moved to larger towns, while other shut down for good. Economic depressions in

1873, 1885, and 1894 abetted this process. One reason stated for the decline of Enfield's population in 1885 was "the dullness of the manufacturing there." [19]

Another problem suffered by the valley also affected much of rural New England; the decline of agriculture. Most farm produce (such as grain and livestock) brought lower prices in the 1870s, due to competition from western markets newly served by railroads. Timber products, used for heating and railroad construction, held their prices longer, but fell by the end of the century due to the decline in railroad tie production and the increased use of coal for heating. [20]

Diversification of production and the change to other kinds of farming took place as profitable uses for farms were sought. The production of dairy goods became popular in the valley, with cheese making in Greenwich and Prescott, and a creamery in Millington. [21]

As agriculture became less profitable for many valley farmers, they sold out and moved, or worked in the factories still operating. A great reduction in the number of agriculturally employed took place between 1870 and 1920. Some towns blessed with rich soil, such as Sunderland and Hadley, saw immigrants moving in to take the place of departing native farmers. This did not take place in the Swift Valley at the turn of the century, due to the poorer soils there. One of every thirteen acres of Enfield farmland had been abandoned by the turn of the century. [22]

During the first few years of the 20th century, the valley dairy farmers tried to survive by switching from producing cheese to cream, then to whole milk (for shipment to nearby larger towns). Tightened health restrictions on milk production brought this to an end, as the new equipment required made it uneconomical for most of the small valley producers. By 1925, only six farms in the valley could support more than two dozen cows. [23] That same year, each of the four towns had several farms raising in excess of 500 chickens, and it became "evident that poultry raising was one of the most important industries in the area." [24] However, the introduction of bulk feed, which again was economical only for large farms, was in the process of making this occupation unprofitable as well.

Similarly, market gardening in the conventional sense was no longer profitable in the Swift Valley. By the 1920s, fresh vegetables could be shipped to the cities throughout the year by such places as far away as California and still be cheaper. Many Connecticut River Valley farmers learned to survive by specializing in certain crops, such as tobacco or onions. The smaller farms and poorer soil in the Swift Valley inhibited those farmers from attempting the same. Some managed to make a living off such specialties as berries, potatoes, apples, and honey. These people were outnumbered by those ekeing out a living on the family farm while working in one of the mills during the colder months. [25]

As conditions in the valley got worse, the population kept declining. Between 1850 and 1890, the four towns lost more than a third of their inhabitants. Another 550 people left the valley in the next thirty years, mostly from the poorer towns of Greenwich and Prescott. Over 2½ times as many valley men fought in the Civil War than in the First World War. In 1920, it was noted that only 29 of the 354 municipalities in the state had not recently gained in population. All four of the Swift River towns were on that list. [26]

The declining population of the Swift Valley did nothing to encourage commercial activity there. In the late 1800s, almost every village within the four towns (averaging three per town) had its own store or trading post. By 1920, this had been reduced to a number of general store-post offices in the major villages, and a few scattered filling stations. Enfield had the only chain store in the valley. Most residents did their shopping in the nearby larger towns of Ware and Athol, or took the train to Springfield. Some of the area stores might have closed sooner had not a small tourist trade developed. [27]

Among the physical signs of backwardness in the area was the lack of paved roads in Prescott and most of Greenwich. Both towns lacked electricity and telephone service over much of their territory. [28] Only two numbered state highways ran through parts of the valley, while the train service was reduced to reflect falling demand. Many of the houses photographed for the Metropolitan Commission inventory were run down and unpainted. Tourists and engineers visiting the valley noted the decrepit houses, barns, yards and overgrown farmlands they saw. [29]

Civic organizations managed to stay alive in the valley towns, but some of the churches had a difficult time due to declining enrollments. The Prescott Congregational Church disbanded in 1924 for this reason. Prescott also lacked a fire department, among other municipal services, due to its small size and poverty. [30]

Ironically, the failure of the valley towns to grow and change resulted in the area becoming increasingly attractive to those living outside of it. The plentiful wildlife populations in the sparsely settled area drew many hunters, while the Swift River (especially the wild West Branch) became one of the more popular trout streams in the state. Others, such as Evelina Gustafson, fell in love with "those lovely little towns with their rolling meadows and lofty hills," and the "quaint" Yankees occupying them. [31] Such a pastoral ideal drew many to build summer homes and cottages on the shores of the valley's numerous ponds. Five summer camps were established in Greenwich and Dana by 1922 for children from cities in the region.

Despite the renewed interest in the valley by some outsiders, it was a case of too little, too late for the future of the area. Other powerful outsiders were casting covetous glances at the valley. The declining rural nature of the towns, the low density of their population, the lack of any major cultural or economic landmarks, and the physio-geographical suitability all combined to make the valley of the Swift River an ideal place for engineers to propose as a reservoir site.

PART II   BOSTON'S HISTORIC NEED FOR WATER

Chapter One   Early Developments

The City of Boston, as first settled in the 1630s, was located on a spit of land shaped like a frying pan. Since little surface water existed there, springs and later driven wells supplied the inhabitants with fresh water.

The first recorded attempt to improve the water supply of the city was made in 1652, when the inhabitants of Conduit Street (now Blackstone Street) petitioned the General Court to create a company to pipe water to them. The resulting "Water Works Company" piped water from nearby springs and wells into a small reservoir about four meters long on either side. side. The pipes, probably made of wood, no longer exist today, although the the conduit was in use for over a century. Both firefighting and domestic uses claimed the water. [32]

In 1795, the next major development in expanding Boston's water supply occurred. On February 17, The Aqueduct Corporation was created to pipe fresh water into the city of 20,000. A gravity supply from Jamaica Pond in Roxbury was transported through four main log pipelines. The General Court and the town of Roxbury were to regulate the price of the water. [33] Fire protection was a main concern in obtaining water, so the current equivalent of hydrants were installed at points along the pipelines. Due to the small capacity of the pond, and it's low elevation, only a relatively few domestic users were served. [34]

By 1822, when Boston was incorporated as a city, it had grown to over 50,000 inhabitants. Three years later, Mayor Josiah Quincy chose a committee to investigate the possibilities of supplying the city with more water. That June, the committee reported that a new supply was needed, and recommended that a study of possible sources be made.

A study made by Professor Daniel Treadwell was made public in September of 1825. Besides coming to some strange conclusions about water consumption, the report noted that Spot Pond, north of the city, would make a good water source. The city council took no action on the report. Uncertaintly existed over the issue of having the city supply the water or allowing private corporations to do it. This question was to plague the city for twenty years.

A second water study committee was appointed in 1832. This group retained the services of Colonel Loami Baldwin, an eminent engineer, to investigate the problem and report on solutions. Colonel Baldwin's 1834 report recommended piping water from Long Pond in Natick, or the Framingham ponds to supply the city. The report also noted that one fourth of Boston's wells were contaminated, accenting the need for a new large source of water. The city council took no action on the report. [35]

The Boston Hydraulic Company was formed by a group of city residents in 1836. The General Court granted the company the right to use the Mystic Lakes and Spot Pond, both north of Boston, for water sources. Up to one thousand shares of stock were to be issued to finance the project. The city of Boston was given the right to subscribe for one third of the stock, but the city council voted not to do so. This left the company underfinanced, and it collapsed.

In 1837, a third study committee was appointed, but it could not agree on which water source was best. A year later, Mayor Eliot filed a petition with the General Court to seek the right to secure a water supply for the city. By 1840, the bill had died because of opposition and lack of support from Eliot's successor.[36]

A second aqueduct corporation was formed in 1843 to bring water from Spot Pond to the city. The city again failed to subscribe to stock in the company. Public agitation forced a referendum on the issue in 1844. The vote was 6,260 in favor of going to Long Pond for water, and 2,204 against.[37] As a result of the vote, a petition was filed in the General Court in 1845, seeking approval for the project. Despite strong opposition, the bill passed that March. The act stipulated that the city must accept the measure in another referendum before it became effective. Intense lobbying by opponents of the project paid off as the voters turned down the act. Proponents tried again in 1846, and succeeded in getting enabling legislation through the General Court, and city approval for the project.[38]

The act created a board of three appointed commissioners to supervise the project. The Commissioners were authorized to fund construction through bond issues. A receiving reservoir for the water was built in Brookline, with distribution mains laid to three Boston storage reservoirs. Long Pond in Natick had its level raised, and was renamed Lake Cochituate. A fourteen mile masonry aqueduct was built to conduct the water from Cochituate to Brookline. The whole project cost the city about $4,000,000.[39]

On October 25, 1848, a great celebration was held in Boston to mark the introduction of Cochituate water into the city. A crowd of 50,000 gathered around the fountain in the Frog Pond at the common. A valve was opened, sending up an eighty foot gusher of water to the delight of the crowd.[40]

The Lake Cochituate supply proved adequate for a little more than twenty years. In 1869 arrangements were made with Charlestown to use part of it's Upper Mystic Lake as a water source. Later Boston aquired the whole lake for its use, but this had to be abandoned in 1898. The Chestnut Hill Reservoir was constructed in 1870 to provide an additional storage reservoir for the city.

A few years later a water shortage threatened the expanding city again. The Sudbury River Act was passed in 1872 to authorize Boston to divert water from that river in Framingham, west of Cochituate. A temporary connection was installed to make use of the water as soon as possible. The Sudbury Aqueduct and three Framingham reservoirs were completed in 1880. No sooner were these works completed than it became necessary to add more sources of water for the system. The Ashland and Hopkinton Reservoirs we were constructed on the South Branch of the Sudbury River. Two more storage reservoirs were added to the system near Boston in the 1880s. [41]

The construction of these aqueducts and reservoirs set a trend in the direction of Boston's future water supply expansion. Each succeeding source of water was found further west of Boston, and came from pure upland lakes and streams. This trend paralleled that of New York City, which went north to tributaries of the Hudson River for its water.

A dry period in the early 1890s brought the water problem into focus again. What had caused the capacity of the metropolitan water supplies to become inadequate? Between 1820 and 1890, Boston quadrupled its land area by filling in the Back Bay and annexing adjoining cities and towns in whole or part. * This brought the area of the city to nearly 40 square miles, and it included some hills up to 200 feet above sea level in the southern parts of the city. This was a wide area to provide adequate water supplies for.

The population rise of the city paralleled its expansion in land area. From 1850 to 1890, Boston averaged a growth of over 70,000 people a decade. Between 1890 and 1895, the population of the city rose by over 50,000, totaling 504,078. At the same time, the population of the whole Boston Metropolitan area rose by 14.8%, which put it over the million mark. [42]

Along with population growth, the corresponding rise in industrial activity added to Boston's water consumption. 5,543 manufacturing establishments were noted in the city in 1895, and this equaled one tenth of the total number of the city's buildings. By 1894, the city's water consumption (domestic and commercial) had passed 90% of the daily capacity of its water souuces. [43] Almost no metering of individual usage was done. and adding to the problem was the common practice of allowing water to run through the tap all night to prevent pipe freeze-ups in cold weather. [44]

The water problem was even more acute in the suburbs around Boston. In 28 municipalities around the city, it was noted that even though water use had decreased between 1890 and 1894, total consumption was running about 2,000,000 gallons ahead of the safe daily yield. [45] Another threat to many water sources was the encroachment of homes and industries upon them.

In 1893, the General Court authorized the State Board of Health to conduct a study of the water needs of the Metropolitan Boston area. The Board was also expected to recommend solutions in the form of new sources of

---

*The corporate units annexed by Boston were Roxbury, Dorchester, Charlestown, Brighton and West Roxbury.

water. The assignment of this board over any local body indicated the regional nature of the water problem, and the ability of the Health Board to handle the investigation.

Chapter Two   The Board of Health and Its Investigation

The Massachusetts Board of Health had been legislated into existence in 1886 to replace an ineffective State Board of Health, Lunacy, and Charities. The new Board was given broad executive powers to enforce state health laws, and to do research on disease prevention and water pollution.

Dr. Henry P. Walcott (1838 - 1932), who had been active in previous state health agencies, was named the first chairman of the new Board. A progressive in his field, Dr. Walcott quickly moved to form a viable organization. An engineering department was created, to "have general oversight and care of inland waters and to advise cities, towns, and other and the State government as to the water supply, drainage, and sewerage." [46]

The three engineers initially hired for this department all had a lasting effect on the Massachusetts water story, especially one. The three were Frederic P. Stearns, Joseph P. Davis, and X. H. Goodnough. Stearns and Goodnough, like Walcott, were Harvard graduates -- proper members of the engineering elite that was just beginning to influence urban growth policies in the U.S. .

Stearns (1851 - 1919) came from Maine. He had spent 14 years investigating water and sewerage problems in the Boston area. He was made chief of the Department of Sanitary Engineering for the Board of Health by Walcott. With his colleagues, Stearns was to work out the plan which still influences the selection of potential water sources for Metropolitan Boston.

Davis (1837 - 1917) became a consulting engineer for the Health Board after a short time in its employ. He also consulted for the cities of Boston and New York. Davis assisted Stearns in compiling the report mentioned above.

X. H. Goodnough had more to do with the creation of Quabbin Reservoir than any other single individual. He worked for the state health agency for 44 years, and from there assisted in the creation of the Wachusett Reservoir, and pushed through the Quabbin project. The man and his life deserve close study.

Xanthus Henry Goodnough * was born in Brookline, Mass. on November 23, 1860. He was one of three sons of Xanthus and Kate (Harley) Goodnough; the father was an engineer. In 1878 Goodnough entered Harvard

---

*Ironically the name Xanthus is also the name of the Greek river god in *The Illiad,* who overflowed his banks in attempting to drown the hero Achilles.

College after graduating from Brookline High School. Following his father's footsteps, he studied engineering, and received an Bachelor of Arts in that subject in 1882. It is an odd note that later in life, Goodnough allowed others to have the impression that he received a doctorate, which was not the case. [47]

After graduation, Goodnough moved west, engaging in "miscellaneous work on the Wabash & Pacific R.R." in 1883 and 1884. [48] The next year, he moved back to the Boston area, working for the Massachusetts Drainage Commission. A big break opened up for Goodnough in 1886, with the health department reorganization. Dr. Walcott hired Goodnough as an assistant engineer on September 3, 1886. Goodnough's gratitude for getting the job is evident by the eulogy he published on the death of Dr. Walcott 46 years later. The young engineer married in 1892, but the couple did not have any children.

In his extensive travels throughout the state while performing his engineering duties, Goodnough did much fishing in streams that attracted him. One of his favorite spots was the West Branch of the Swift River.[49] Undoubtly while fishing there, he also noticed the possibilities of the Swift Valley as a reservoir site.

When the legislature authorized the Board of Health to study water supply problems in 1893, Walcott and the Board put together its in-house team to do the job. Frederic Stearns was named chief engineer, with Joseph Davis as consulting engineer. X. H. Goodnough was one of the assistant engineers. A number of men who had worked for or assisted on earlier Boston water projects also were called in for the investigation. About $37,000 was spent on salaries for these people and their assistants.

The investigators considered all possible water supply sources in New England. Sebago Lake in Maine and Lake Winnipesaukee in New Hampshire were rejected for their distance from Boston and out of state ownership. The Merrimac River was felt to be too polluted to use (although it was acceptable for the City of Lawrence!). The Charles River was judged to be not clean enough, and susceptible to too much future population growth in its watershed. The Deerfield River, mentioned as early as 1892, was rejected on account of its distance from Boston. [50]

This process of elimination left the South Branch of the Nashua River as the choice of the Board for a water source. The valley of that river above the Lancaster Mills in Clinton was a basin formed by the glacial Lake Nashua. A sufficient depth could be achieved for the proposed reservoir, and it would be high enough above sea level to feed the water works to the east by gravity. Consideration of what the reservoir would displace was not felt to be a great hinderance:

> ... we have been impressed by the very serious changes which will be produced in the towns of Boylston and West Boylston. It does not appear to us to be a very important objection to our plan that certain mill sites will be 80 feet below the surface of the basin, nor that the homes of many industrious people dependent upon these mills for their living will also be submerged, because all these can be paid for, and an equivalent will be given ... [51]

Over half of the assessed value of West Boylston, and over a third of Boylston was to be taken for reservoir purposes.

The Nashua River was not only a good supply in itself, but it was conveniently located for a spring board to further extensions to the system. Other streams in east-central Massachusetts, such as the Assabet River, were considered useful as future water sources. But the Nashua River was further west, and pointed the way for greedy eyes to other streams:

> The very great merit of the plan submitted is to be found in the fact that the extension of the chain in the Metropolitan water supplies to the valley of the Nashua River would settle forever the policy of the District, for a comparatively inexpensive conduit can be constructed through to the valley of the Ware River and beyond the Ware River lies the valley of the Swift. [52]

Plans for a small diversion reservoir on the Ware River at the Oakham village of Coldbrook were spelled out. A huge reservoir to fill the Swift River also was proposed. Although left in rudimentary planning stages, the proposals would be unearthed at a later date and fleshed out. While the destruction of most of two villages were eminent, the seeds for dismantling four towns had been planted.

The Board filed its voluminous report as House Document No. 500 in February, 1895. Of the 26 men on the two committees considering the concerns of the towns to be affected by the proposed reservoir, only two came from Western Massachusetts. Hearings on the bill were held from February through April. Besides considering the concerns of the towns to be affected by the plan, testimony from Metropolitan towns also was taken by the committees. Part of the bill provided for the formation of a Metropolitan Water Board to oversee the water supplies of all towns within ten miles of the State House.

It appears that the towns which would suffer most from the project accepted their fate and lobbied for adequate compensation for those dislocated by it. About a month after the hearings ended, Chapter 488 of the Acts of 1895 was passed on June 5. The 27 page law set up a three man Metropolitan Water Board, to be appointed by the Governor. The 13 municipalities comprising the district had little say over the board's policies, but they had to foot the bill for its work. The board was authorized to raise up to $27,000,000 by bond issues to finance the project. Besides provisions for settling disagreements between property owners and the board, businesses in the affected area were given the right to seek compensation for loss of their established trade or customers.

Chapter Three   Wachusett Reservoir

In the 1890s, Boylston was viewed as one of the more prosperous agricultural towns in Worcester County. The town had an assessed valuation of about a half million dollars; a quarter of that was the value of farm produce. The only major manufacturing done in the 110 year old town was at Sawyer's Mills, where a cotton mill subsidiary of Clinton's Lancaster Mills operated. The construction of the reservoir would force the removal of 302 of the town's 770 people, a church, and two schools.

At the same time, West Boylston, created in 1808, stood to lose much more. The town's assessed valuation was close to a million dollars, and its manufacturing industries were growing. The reservoir would force out a third of the town's dwellings with 43 percent of the population on 870 acres of land. Three churches, four schools, and six mills would have to move or close down. 60 percent of the assessed value of the town would be lost to the reservoir.[53]

Sterling, at the upper end of the reservoir basin, stood to lose only one percent of its people, as would the mill town of Clinton, the location of the dam. Over eight miles of the Central Massachusetts R.R. would have to be relocated on the north side of the reservoir. 19 miles of roads would be abandoned, and new ones around the reservoir had to be built.

The Metropolitan Water Board was formed and began to select personnel to carry out the reservoir project. Henry Sprague, a Harvard graduate and noted politician, was the board chairman, Wilmer Evans and John R. Freeman, a noted engineer, were the other two members. The chief engineer selected was Frederic Stearns, who had headed the study group proposing the reservoir.

Surveys and land purchases began within the first year of the board's operation. Settlements had to be made with New Hampshire interests, as the Nashua emptied into the Merrimac River at Nashua, New Hampshire.

At the same time, residents of West Boylston realized that the closing of local mills might leave them without a job for a period of time until they could relocate. The General Court was asked to remedy this problem. Legislation was passed to allow West Boylston employees to file for compensation no greater than the sum of their wages for the last six months of employment.

Much of the labor hired to do the physical work on the project consisted of Hungarian and Italian immigrants. These nationalities were new to the Boylstons, and some of their habits were disturbing to the natives. A strike occurred, with some attendant violence, in 1899, when crew bosses tried to cut wages. The conflict was settled after a few days. [54] In the peak year of 1901, over 3,000 men (and 324 horses) were employed in the work force.

In 1897, agents from other towns seeking industries began to visit West Boylston. Two of the major mills in town relocated after selling out to the state, taking many employees with them. The others shut down for good by 1902. Several hundred people ended up leaving Boylston, while West Boylston lost half of its residents. 447 claims for unemployment compensation were filed, and $84,959.65 was paid out for that purpose. [55]

Work on the dam began in 1901, while soil was being stripped off the 6.5 square mile basin to be flooded. Buildings were demolished, but not before some of these had been rented to members of the work force for a while.

That same year, the Metropolitan Water Board was combined with the sewerage board to become the Metropolitan Water and Sewerage Board. Henry Sprague remained as the only member from the water board. Dr. Walcott returned to the scene as a new member, and a young Harvard educated engineer named James A. Bailey, Jr. was the third member. This new board was to last two decades, overseeing the completion and later depletion of the Wachusett Reservoir.

Contract work on the reservoir continued until 1906. On May 10, 1908, the 64 billion gallon capacity of the waterworks was reached. The project had cost just over $11,000,000, including $3,700,000 in real estate costs and damages.

In the same period that the Wachusett was being completed, other great water projects were being proposed or carried out around the nation. In 1908, the "conservationist" Roosevelt administration gave the city of San Francisco permission to flood the magnificient Hetch-Hetchy Valley in Yosemite National Park, for a water supply reservoir. The city had other available water sources, but contended it was cheapest to go to the Hetch-Hetchy. Conservationist John Muir, in opposing the project, spoke of the engineers behind it as

> ... the spoiler(s) ... These temple destroyers, devotees of ravaging commercialism, seem to have a perfect contempt for Nature, and instead of lifting their eyes to the mountains, lift them to dams ... [56]

The project went through anyway, to the pride of those who ended up constructing one of the longest aqueducts in the world.

While the Wachusett Reservoir was under construction, New York City was preparing to expand its water supplies. The Croton watershed, near the Connecticut border, had been developed to its capacity. In 1902, a commission was appointed to find solutions to the problem. One of its members, John Freeman, recently had served Metropolitan Boston as a water advisor. The commission decided that a good source of water for New York would be in Ulster County, west of Kingston in the Catskill Mountains. The Scoharie and Esopus Creeks would be dammed to create large reservoirs. A long aqueduct would connect these waters to the Croton system. Preparations for the project commenced in 1906.

Eight villages were destroyed to create Ashokan Reservoir, which held the waters of Esopus Creek. The city spent $188,000,000 to complete the project, including $17,300,000 to purchase 21,330 acres of land. The rapaciousness of the land aquisitions was compared to Metropolitan Boston's "predatory methods for aquiring property," for Wachusett. [57] After all of the court suits were cleared, legal costs incurred by New York City amounted to 63 cents for each dollar awarded to claimants in regard to the project. One of the lawyers from nearby Kingston, N.Y. who was engaged in water cases was a John D. Schoonmaker, whose Ware, Mass. relative would later represent the foes of the Quabbin project.

It may be useful at this point to note some of the subtle shifts of power that had taken place in politics and society, which ultimately would affect water supply considerations.

The rise of urban commerical and industrial centers in the last half of the 19th century shifted the balance of political power in places such as Massachusetts. Boston, which had been a big brother among equals in the early part of the century, now represented the dominant part of the state. Its own population had grown to the point where referendums, such as the water votes taken in the 1840s, were not viewed as being practical. The city gained power, but the power was wielded more often by bosses and bureaucrats, ostensibly representing the inhabitants. The small town, and its ways lost out to the overwhelming power and pervasive influence of cities.

As the bureaucrats became entrenched, they formed elites. Since they possessed the training to do their jobs, and did not often have to answer to the public, their "knowledge was power, privileged power."[58] The new technological elite would seek the consent of the governed, through their legislators, but clearly they would govern, not the people.

Engineers were one of these bureaucratic elites. As in the case of Massachusetts, most of the engineers in government entered through the expanded public health programs of the late 1800s. The administration set up for the program, with its rational data and technical assistants, gained the power once held by the rotating boards and commissions of an earlier time. [59]
The needs of a metropolitan area, such as water supply, were taken advantage of by the bureaucrats to formulate grandiose schemes for solving them.

If the scheme went through, jobs for friends and contracts for businessmen were available.

Massachusetts saw a Harvard elite, and later a Brown University elite move in to plan and operate these engineering projects. Many of those running the projects hired friendly "outside" consultants, who in turn hired their benefactors to consult upon out-of state projects. Men such as Stearns, Freeman, and Goodnough were not unusual in gathering the fruits of merry-go-round jobs. The question, which still exists, is whether these engineers were really serving the needs of the people around them, or just creating greater jobs and glories for themselves.

Chapter Four   More Cities, More Water

When the Report of the Board of Health in 1895 mentioned the Swift River-River Valley as a possible source of water for Metropolitan Boston, it received front page notice in a local paper. * The reaction of valley residents was a flurry of consternation, but most dismissed the project as an impossibility.[60] A few people came from the basin of the Nashua River and settled in the valley, apparently confident that they had located a permanent home there.

In 1899, a proposal was made by the Springfield Water Co. to flood the West Branch valley between Pelham and Prescott to create a water supply for Springfield. The jobs the project would generate and the low value of the land in question were seen as making the proposition a reasonable one for the area. Nothing came of this proposal, however.[61]

In 1909, a semi-secret investigation of the Swift Valley was made by engineers for either the Board of Health or the Metropolitan Board. No public document was ever issued on the study. Newspapers in Worcester and the valley area made note of the presence of the engineers and recited the plans revealed in 1895. It was stated that the proposal would merit serious consideration within ten to fifteen years. A North Dana news correspondent optimistically felt "it is safe to say the day is far distant when it will be done. North Dana people don't need to move before snow flies, at any rate." [62]

Five years later, the First World War burst upon the international scene. The demands of the war increased industrial production even before the United States joined the fray. In 1916, the City of Worcester began to show concern for its water supply, and petitioned the state to conduct an investigation of the problem. At the same time, the Metropolitan Water and Sewerage Board also became concerned about the capacity of its system, as increased use and lower rainfalls were depleting its supplies.

Worcester was the first to take action. In 1918, it got the General Court

---

*The Ware River News

to pass Chapter 178 of the Special Acts, which set up a study commission to recommend a solution to Worcester's dwindling water supplies. That same year, the report of the Commission on Waterways and Public Lands on the state's water resources stated its opinion that the Ware River would be needed for metropolitan water supply within a very short time.

The report of the Department of Public Health (the "Board" name was changed on 1915) for 1918 included a report by X. H. Goodnough, now chief sanitary engineer for the department. Goodnough strongly recommended a new water study be undertaken by his department and the Metropolitan Water Board. He felt that the Metropolitan Boston system, and other water systems in Eastern Massachusetts, were beginning to overrun capacity. He pointed out that the 1895 report had projected that the Metropolitan Boston system would need re-evaluation in twenty years, which had already been exceeded.

In 1919, legislation was filed by the Health Department and the Metropolitan Water and Sewerage Board to authorize and fund a new statewide water study. In June, Chapter 49 of the Resolves gave the General Court's blessing, with no particular limit on spending. A joint board consisting of members and staffs of both departments was to carry out the study, to be completed in 1921. This was later extended to 1922, at the request of the Joint Board.

How did it come to pass that Boston and its suburbs had outstripped their water supply in only a dozen years? Between 1890 and 1920, Boston had annexed only one small area, known as Hyde Park. While the land area of the city did not increase substantially in these years, its population did rise by over 50%, to 748,000. Likewise, the cities and towns in the Metropolitan Water District grew by 60% in the same period, to a total of 1,206,849.

It is curious to note that the estimate of the engineers in 1895 of the population increase for this period was high by 700,000 people. The water consumption also was overestimated by 70,000,000 gallons a day. True, water consumption had been higher during the First World War, but it fell off somewhat afterward. The water district used 127,000,000 gallons a day in 1920, and it was conservatively estimated that at the present rate of increased use, the capacity of the system would be outstripped by 1930.[63]

One of the causes for concern about the Metropolitan water supply was the increase the city of Worcester was making on it. In 1920, Worcester was purchasing about five million gallons a day from the Wachusett supply. The study commissioned in 1918 reported two years later that another reservoir for Worcester could be built on a tributary of the Wachusett, and that the Ware River could be a future source of water. This conflicted with present and future sources of Metropolitan District water. It was becoming obvious that Worcester and the District would have to share sources in the future, but where, how, and at what cost would have to be determined.

A final consideration in the water supply problem was the needs of cities and towns eligible to join the Metropolitan Water District, but not yet within it. According to the 1895 legislation, any municipality within a ten mile radius of the State House could join the district. The total consumption of these towns and cities had increased 75 percent over the period from 1890 to 1920, and many had reached their capacity. [64] Other places which still had a good water supply were faced with increasing encroachments upon the water sources by the growth of population.

The Joint Board was organized on July 3, 1919. Eighty-one year old Dr. Walcott, still chairman of the Metropolitan Water and Sewerage Board, also chaired the new board. X. H. Goodnough was the secretary for the board and engineer, while Frederic P. Stearns became a consulting engineer. This familiar cast of characters was at the head of a group of relative non-entities from the metropolitan area, including Harvard friends of Goodnough. The only exception was engineer James Bailey, who had served on the Metropolitan Board with Dr. Walcott since the turn of the century.

Some of the personnel hired by the Joint Board obivously reflected the in-house viewpoint. After Frederic Stearns died in December, 1919, he was replaced by J. Waldo Smith, who had worked on Massachusetts water problems when Goodnough was still new in the Health Department. Charles T. Main, a legal consultant, had done the same job for the 1895 study. A few of Goodnough's assistants at the Health Department (including Karl Kennison) also worked for him under the Joint Board.

Besides the addition of J. Waldo Smith, the board underwent other changes by the end of 1919. Dr. Walcott resigned as Chairman in December but stayed on as a consultant. Dr. Eugene Kelley, who as Commissioner of Public Health was Goodnough's boss, replaced Walcott as board chairman The newly created Metropolitan District Commission replaced the old water and sewerage board as one of the two "joint boards". James Bailey continued as a member of the new Commission and the Joint Board.

Goodnough must have been formulating a Swift-Ware proposal for Metropolitan water expansion right through this shakedown period. Though he stated that "little had been accomplished" by the end of 1919, [65] newspapers ran stories on the plight of the threatened Swift River towns late in 1919. [66] The attention his proposal received also is borne out by the fact that in a report that was meant to cover the whole state, areas outside of Metropolitan Boston and Worcester were considered in only one-thirteenth of the report.

The Joint Board had been given two years (until 1921) to submit its report by the enabling legislation. Due to "unexpected delays" [67] never adequately explained, the board asked the General Court for another year to complete its study. This caused a furor among many people opposing the

Swift-Ware proposals. Many felt that further delays in resolving the question would only hurt those living in potential project sites. [68] Town meetings in the Ware and Swift Valleys passed resolutions opposing the project and retained counsel to represent them at legislative hearings.

Some thought the proposals preposterous and impossible to carry out. Symbolic of this attitude was a comment in an Athol newspaper.

> What a fine place for boating that big Metropolitan reservoir will supply, and Old New Salem from her lofty and breezy heights will be able to enjoy a water view of almost oceanic grandeur! Big ships may sail from the piers near South Athol for a voyage South almost to the borders of Three Rivers! * Wonderful things may happen if this stupendous scheme materializes. [69]

Hearings were held by the General Court in mid-March on the bill to extend the limit for the Joint Board report. Eugene Kelley and Goodnough began by stating the needs of the Metropolitan area for a better water supply. They also felt they needed more time to complete a proper study. Opponents argued against taking the Ware and Swift waters, and that more time should be given for the investigation to exhaust all possible alternatives. After the hearing, Goodnough button-holed some of the milder opponents, assuring them that he would give them any information they needed about the projects. [70]

A couple of weeks after the hearings, the representatives took up the matter of the extension bill. Much acrimonious debate ensued about the voracious character of the "Metropolitan Hygiene Hierarchy." Rep. George P. Webster of Boxford, a town well north of Boston, led the attacks on the State Department of Public Health. It was noted that the Health Department was trying to use the water needs of Lowell, Lawrence, and Haverhill as part of the justification for expanding the Metropolitan system. Representatives of three Merrimac Valley cities stated that in spite of statements by the Health Department they did not want to join the Metropolitan Water District.[71]

Webster stated his feeling that all water supplies would eventually become filtered, and that the Health Department's motive for seeking only pure sources was political. He also opposed what he felt would be the enormous cost of the Swift-Ware projects.

Restating these views a couple of months later in a speech, Webster added that X. H. Goodnough was "so smooth that he could sell Christmas trees in a synagogue." [72]

Amendments were added to the bill, including one requiring the investigation to include a study of using filtered water from the Merrimac River.

---

*Three Rivers is a village of Palmer, Massachusetts.

Another amendment added consideration of artesian wells to the study. The bill was finally approved in early May, 1921. The Joint Board had another seven months to complete its report.

The waiting period on the report was marked by continued expressions of opposition by residents of the Swift and Ware Valleys. It also saw the visit of a joint legislative committee to Enfield, to see where the potential reservoir sites were located. [73]

Chapter Five   The Joint Board Report

On January 28, 1922, the Joint Board filed its 284 page report with the General Court. Fourteen specific recommendations were made by a majority of the board. The most important were:

1. Definite arrangements should at once be made for securing an additional supply of water from the Ware River for the joint use of the Metropolitan Water District and the city of Worcester.
2. A tunnel about 13 feet in diameter and about 12 miles long should be constructed from the Wachusett Reservoir westward to Coldbrook on the Ware River at an estimated cost of about $12,000,000.
3. This tunnel should be used to divert the flood flows of the Ware River to the Wachusett Reservoir, securing an additional supply of water of about 33,000,000 gallons per day, in such a way as to interfere as little as possible with the mills on the lower Ware River.
5. The Swift River project, involving the construction of a great reservoir near Enfield, and connecting it with the Wachusett Reservoir by means of an extension of the tunnel to the Ware River, should be approved as the logical extension of the Water supply system.
7. ... keeping in mind the ultimate filtration of all of the metropolitan water supply.
9. Efforts to prevent leakage and waste of water should be continued in order to conserve the present supply to the greatest possible extent.
10. For the same reason the local supplies ... should be developed to their economical limit ... keeping in mind the possibility of their ultimate abandonment.
11. The construction of the proposed works should be entrusted to the Metropolitan District Commission or to a special commission, as the Legislature may determine. [74]

Obviously under Goodnough's influence, a majority of the Joint Board rejected supplying the Boston area from sources east of Worcester. While conservation and extension of local supplies were given encouragement, the "ultimate abandonment" of these supplies would be made that much easier by the construction of a "great reservoir." Goodnough sold the "logic" of his Swift

River project to a majority of the board, as it is the first thing mentioned after the Ware Project

One member of the Joint Board disagreed with the conclusions of Goodnough and the others. This was James Bailey, of the Metropolitan District Commission, who filed a ten page minority report.

Bailey noted that population increases within the Metropolitan District did not match figures projected in 1895. He added that the population increases would be even smaller due to lower birth rates and restricted immigration. This led him to believe that population increases alone would not cause a need for more water sources for another decade.

Recognizing that Worcester was in need of water, Bailey felt that the Metropolitan District and that city should share water from the Ware River. He was opposed to the Swift River project, believing that present District supplies plus the Ware River would be adequate. [75]

Other engineers connected with the board seconded Goodnough's recommendations, especially J. Waldo Smith, an old friend of the former. Smith's consulting engineering report was only a five-page warning that something to augment the Metropolitan supplies be done as soon as possible.

As chief engineer for the board, Goodnough wrote the main part of its report. Marshalling an impressive array of statistics, he tried to indicate an immediate need for a larger water supply. As in the 1895 report, what later proved to be inflated population projections were used. While Goodnough correctly predicted that more towns would want to join the Metropolitan District in the near future, he thought they would join more quickly then they did.

Concerning sources for new water supplies, Goodnough's report graciously conceeded that areas south and northeast of Boston should have their streams for their own water needs. While on one hand he consigned the Westfield River to the needs of the Springfield area, he still kept it in reserve for the Metropolitan District. Goodnough rejected such far off sources as Sebago Lake in Maine and Lake Winepesaukee in New Hampshire, but this had been done in 1895.

Goodnough rejected the Merrimac River as a source for the Metropolitan area, again following the lead of the 1895 report. Although in other parts of the report he saw that future Metropolitan supplies would be filtered, he did not want to use Merrimac water because it would have to be filtered. What was good enough for towns along that river was not good enough for the Metropolitan District.

The Concord and Assabet Rivers, located between the Wachusett and Sudbury systems, were rejected as possible sources for rather weak reasons. True, the rivers were suitable only for a series of small storage reservoirs, but this had been Boston's policy until the Wachusett Reservoir was built. The practicability of the rivers as supplies would come up in debates in in succeeding years.

In his ultimate vision for filling the Swift River Reservoir, Goodnough saw over 1000 square miles of watersheds being taken over. Waters of the Ware, Swift, and Millers Rivers initially would be used to fill the reservoir, with the Quaboag, Deerfield, and Westfield Rivers eventually joining in.

The total cost of the Ware and Swift projects, and the tunnel to the Wachusett Reservoir was put at $59,946,540. This figure included the cost of real estate and damages for the taking of water rights. While this figure was not very far below the final cost of the project federal aid and lower depression prices and wages make one suspect it would have ended up much higher under normal econonic circumstances. Voters in the Metropolitan District would balk at this figure because it was half-again as large as the total debt incurred by the Metropolitan Water Projects from 1895-1920. [76]

The Joint Board report was referred to the joint legislative committee on water supply for consideration. There the water fight would truly begin.

PART III   THE WATER FIGHT

Chapter One   The Politics

The system of government in Massachusetts in the 1920's was basically the same as that set up in the pioneering state constitution of 1780. As in most republican governments, there were three branches; executive, legislative, and judicial.

The executive had six elected officers; the Governor and Lieutenant Governor, Attorney General, Treasurer, Auditor and Secretary of State. The court system, under a State Supreme Court, consisted of district courts tied in with the counties. A curious anacronism (which still exists) in the executive branch is the elected Governor's Council. This quasi-judicial body approves some of the governor's appointments and pardon requests.

Most important to this story is the legislative branch, which goes by the formal title of the Great and General Court. In the 1920's, this consisted of a 240 member house and a 40 member Senate. As with most state offices since the Civil War, the Republican party dominated both houses of the legislative branch.

Geographically, the largest block of legislators (about one third) lived within the Metropolitan District and its immediate suburbs. While not as large a proportion as it would become later, this voting block was not as unified either. Most of the minority Democrats came from Boston districts, while the suburbs and rural areas were largely Republican. [77] Still, with the State House and most of the dominating business interests of the state in or near Boston, it s influence was certainly stronger than any other single area within the state.

Possibly due to the strong Republican makeup of the house, it was dominated by older men. Most members in 1922 were 45 or older; many were retired executives or businessmen. [78] The General Court was just a springboard to higher office for some; Calvin Coolidge was just one of the recent governors of the state produced by it.

The early 1920's editions of the General Court had such future luminaries as John McCormack, Leverett Saltonstall, Frank Allen (later Governor) and Elijah Adlow (later a noted judge) in its ranks. However, men such as the high-brow Henry Shattuck, a Boston lawyer, actually had the most power at the time. [79]

Legislating was not the fulltime job it became years later. Most sessions of the General Court were finished for the year by June, and one could not live very well on a legislator's salary. People were closer to their lawmakers, as each one represented only about 15,000 people, and less time was spent at the State House than in later years.

The Metropolitan District Commission and the Department of Public Health were both parts of the executive branch of the state government.

Both reflected the rest of state government in that they were staffed by members of the Republican professional class. Curiously enough, most of the constituenry the Metropolitan District Commission served was made up of Catholic-Democratic immigrant groups. Even though the engineers and planners might not be politically allied with their constituents, they could further their own careers by developing schemes which seemed to serve the people best.

Men such as Goodnough best exemplify the dichotomy of the citizens and their servants in the executive branch of government. A close look at some of the key legislators involved in the five year fight for the Swift River Reservoir will also be instructive.

One man from the Swift-Ware region served in the lower branch of the General Court from 1913-1940. He was Roland D. Sawyer, a Congregational minister from Ware. A rather unique man for his time, he was not averse to an occasional drink even though he was a Protestant minister in the days of Prohibition. [80]

Born in Kensington, N.H., in 1874, Sawyer began his ministry after his marriage in 1898. He preached in Brockton, Hanson and Haverill, Mass. before settling in Ware for a long stay after 1909. Sawyer was a political activist as soon as he could vote. The Anti-Profanity League of Teddy Roosevelt's era was founded by him, and he was an active "Hearst" Democrat in the same period.

After moving to heavily ethnic Ware, Sawyer became active in the State Socialist Party. He was the party's candidate for Governor in 1912, but left it in a dispute the next year. Rejoining the Democratic Party, he was elected to go to the General Court from a district including Ware, Enfield, Greenwich, and Prescott. Although the Republican Swift Valley towns never gave him a majority of their votes, his hometown of Ware was heavily Democratic and the largest town in the district.

Taking up the cause against the Metropolitan water schemes to protect the interests of his constituents, Sawyer became the most consistent critic of them in the next five years. Sharply contrasted with him were his collegues who filled the seat representing Dana and Athol, especially Leslie Haskins.

Almond Smith, a lumber dealer from Athol, represented that town and Dana (among others) in the First Worcester District from 1920-24. He kept a very low profile on the water issue during his stay in the house. His few recorded statements on the issue were more concerned with keeping the Millers River and Athol out of the proposed reservoir watershed. While this is understandable from a political viewpoint (Athol having twice the number of voters than the rest of the district), it shows Smith to have been too provincial even for his own district.

Leslie Haskins, who succeeded Smith in the house seat, was born in North Dana in 1880. He had been active in local politics while selling insurance and serving as the local railroad agent. He ran for the house seat in 1924, when it was felt that the time was crucial for a Dana man to represent the district.[81] As a Republican in a heavily Republican district, he had no real problem getting elected.

Haskins was given a seat on the Legislative Committee on Water Supply, which was a nice gesture toward the freshman member. His membership in the majority party and his tendency not to speak strongly about the water issue undoubtedly got him the position. He ended up working "for a good bill" as opposed to working against the project.[82] While this can be attributed in part to advice he sought, his favorable impression of X.H. Goodnough and friendship with the Governor would indicate a good party man who went along to get along.[83]

The last legislator we shall note here was George P. Webster, of West Boxford. Born in 1877, he had a varied career as a farmer, salesman, and journalist, as well as serving in the Spanish-American War. He was first elected to the General Court in 1912, as a Progressive. He served two terms, and was just barely defeated in a contest to be elected speaker. Webster's appealing progressivism and "great eloquence" both contributed toward furthering his political career.[84] After losing a bid for re-election in 1914, Webster served in the constitutional convention of 1917-19. He was elected to the house again in 1918, and served there until his death in May, 1923.

Webster served on many committees while in the house, and was chairman of some of them. Even though the progressive movement had faded during the First World War, Webster still advocated women's rights, contributory pension plans, labor rights and fiscal reorganization of the state.[85]

Although not a member of the water supply committee, and not representing either Metropolitan Boston or the Swift-Ware valleys, Webster was the most vociferous early opponent of the water projects. Perhaps a sense of fair play entered into it, but a running feud with the Department of Public Health and X.H. Goodnough was the main cause of Webster's attacks.

Such a volatile issue as a huge water supply project, with its attendant social and economic dislocations, was sure to create a veritable war when thrown into an arena of these personailities.

Chapter Two   The General Court Tries Its Hand

The Joint Legislative Committee on Water Supply received the report of the Joint Board in January, 1922, but it didn't begin to hold hearings on it for several weeks. During this time, much agitation against the report came from the Swift-Ware valley towns. Town meetings were either hiring counsels to represent them in Boston, or asking the Selectmen to do that job.

Concern about the proposals also came from areas outside of the Swift and Ware valleys. Athol and other towns along the Millers River did not want to see any of that stream diverted into the proposed Swift Reservoir. Representatives of the towns and millowners along the Millers went to Boston in early February. After talking to legislators and other public officials, they prophetically concluded that no final action would be taken on the issue that year. [86] Some Bostonians opposed the project feeling that it was too ambitious and costly. [87] James Bailey's minority report was used by these people as ammunition.

The first public hearing of the legislative committee was held in late March 1922. The highlight of the session was James Bailey's explanation of his minority report. He did not think a separate commission was necessary to do the job; his Metropolitan District Commission could do the work. Bailey saw no need for most of Goodnough's proposed water works, feeling that the current slack period in local industry would not raise water needs soon. Filtration of more sources of water close to Boston was Bailey's answer to any immediate water needs, with an eventual hookup (with Worcester) to Ware River flood waters. He also asserted that he got the Joint Board to change legislation proposed in the report which would have called for immediate land takings in any proposed reservoir areas. [88]

Goodnough still pushed for his plans, and tried to make them sound more palatable by reasoning that to avoid building reservoirs now would mean eventual taking of all water in the Swift and Ware Rivers, not just flood flows.

Bailey's remarks, plus the inactivity of the committee for over a month gave some a false sense of security on the issue. A Western Massachusetts newspaper, while noting the overflow at Wachusett Reservoir, joked about big steamships traversing the "dead waste of swirling flood". It also observed "How those who come after us 15 or 20 years hence will laugh as they read of the great schemes for Boston water supply." [89]

Despite such optimism, the lack of a definite yes or no on the matter was beginning to wear down opposition in the river valleys. When the legislative committee held a hearing in Enfield early in May, over 400 area residents showed up. They heard Goodnough explain the details of his plan. Selectmen J.H. Johnson of Dana presented the committee with a petition from the selectmen of the four Swift towns. The document requested that if the project went through, any land required for it be taken by purchase, not eminent domain. Johnson stated, "We are the victims of an unfortunate necessity, This must be a necessity or a bunco game, and I choose to think it is a necessity". [90]

Other speakers at the hearing also asked for prompt action on the part of the state, as well as opposing the project. A public forum held in Ware shortly after the Enfield hearing saw much condemnation of the water projects, led by Roland Sawyer.

When the scene shifted back to Boston on May 19th, things really began to heat up. This time the legislative committee took testimony from a variety

of speakers in and out of government. George P. Webster strongly attacked the Department of Health, stating that "in purely health matters, it is the peer of any body in the world", but in its engineering schemes, it "assumes extra-constitutional powers". Webster thought the department was "concerned chiefly with its own aggrandizement . . . its chief aim is to build up a tremendous political machine".

The expense and dislocation the projects would cause also bothered Webster; "The valuation of the towns to be wiped out, two million dollars, is looked upon as mere cigarette money by the sponsors". He paralleled Bailey's theme that industry was hurting in the state and asked how many other industries would close or leave the state due to the projects. Webster asked how, during a housing shortage, 1400 homes could be leveled.

Dr. Eugene Kelley, the Commissioner of Public Health and Goodnough's superior, hotly replied to Webster's charges. He asked how Webster could praise him on the one hand and come "very close to slander" on the other. He said the special commission called for in the bill could not allow him to create any jobs, and denied that the department had "any interest in carrying out the 1895 report". This latter statement was a poor defense of charges that Goodnough had set the course for what he was now proposing two dozen years earlier. Kelley also scored proposals to use wells and filtered sources of water as inadequate for the future needs of the Metropolitan system.

Other speakers representing the river towns opposed the projects. Representative Pond of Greenfield asked that the Millers River be kept out of consideration for at least a year. E.E. Hobson, counsel for Palmer industries, compared the bill to the ravishment of Belgium by the Germans in the First World War. Hobson stated the bill "removed the living from their homes, disturbs the dead, and disintegrated towns". He wanted more consideration given to the matter, because the decision to go ahead would be irrevocable.

J.W. Wheelwright, a Ware River millowner, was opposed to taking any Ware River water, as it would disrupt industries along the banks. He was supported by a representative of the Springfield Chamber of Commerce, who worried about the effect of losing both Ware and Swift water on his city's industries. J.H. Schoonmaker of Ware felt that his part of the state would be underrepresented on the commission proposed to oversee the projects. He also thought the method of assessing land needed for reservoir purposes was poorly spelled out in the bill. 91

This last statement again indicated a growing resignation on the part of valley people that the project would go through. With this attitude came a concern to get the best deal possible when it came time to sell out. The main topic of the next hearing, on May 23rd, turned out to be the hardships caused by the delay in a final decision and the desire for a good deal. Twenty-five selectmen and leading citizens of the valley towns came to Boston to express this view. They still opposed the project, and felt its cost estimates and surveys had been inadequate

Selectman Johnson of Dana was again a spokesman for the group. He pointed out that business and real estate trade were at a standstill in the four towns while the situation was in doubt. He thought that many residents of the valley would accept leaving if they got a fair price for their properties.

Although a fellow Democrat, Johnson went on to criticize Roland Sawyer for "playing politics at the expense of the small Republican towns in his district". He felt that Sawyer's tactics only served to keep the valley towns in suspense. [92]

Sawyer, who had not been at the hearing, later commented that "any man who will come down here 75 miles to advocate the flooding of the homes of his neighbors, in order that he may make a profit . . . is a traitor or a coward". He scored his opponents for not speaking in his presence, and said the issue was too serious for "political considerations".

Tossing in a shot at Goodnough, Sawyer added that his opponents helped "certain interests about the state house that have long sought to create the impression here that most of the 2800 people to be ousted really want to go". [93]

The next day, several valley residents who attended the hearing answered Sawyer's statement by reaffirming their opposition to the project and their desire for immediate action one way or another. They thought Sawyer's "assertion of ulterior motives is too absurd for consideration".

That same day, the legislative committee reported to the house and senate that the whole matter should be referred to the next annual session. Sawyer tried to have a special commission set up to study the issue in the next few months. He used the ravishment of Belguim arguement in his speech, and referred to Goodnough's Plan as a "wildcat scheme". His arguments failed to sway the house, which overwhelmingly defeated his motion. Another attempt to get his motion through failed - - even George Webster voted against it. [94]

As the session was winding toward a close in early June, Senator Griswold, of Greenfield, succeeded in getting the Senate to approve a bill similar to Sawyer's, but the House would not go along with it. Representative Leo Hamburger of Boston, Chairman of the water supply committee and a supporter of Goodnough, told the house that Griswold was only trying to foster his own ambitions with the bill. Sawyer charged that the feud between James Bailey and Goodnough was spilling over into the General Court, with the Senate supporting Bailey, and the House backing Goodnough. [95]

A few days later, a bill to appropriate $21,000 for an investigation of water supply problems in the northeastern part of the state caused the most vicious debate of the year. Representative Webster opposed the bill, and used the debate on it to attack Goodnough in the strongest terms yet.

Webster began his attack by charging that the Joint Board investigation was "carried out with a view to enhancing (Goodnough's) own political fortunes", Webster added "Falsehood is his habit", asserting that former Governor Coolidge had unsuccessfully demanded that the Department of Health remove him. This occured in connection with Goodnough's supposedly misrepresenting conditions at Lake Cochituate when it was proposed to open it to fishing.

In continuing his attack, Webster stated that "Goodnough wants this $21,000 to spend down in Essex County for the purpose of making friends for himself and enemies for those who would dare oppose him, in accordance with his usual custom. During the past two years he has been making a similar investigation in the western part of the state. He has gone into certain towns there, employed men who had no familiarization for the work in hand, paid them twice as much as would have been necessary to engage competent men, and has mot required them to present itemized bills for their services. In general, he has conducted himself to lead people to believe that he is the Christmas tree for the commonwealth".

Representative Hamburger warmly defended Goodnough in a reply to Webster, getting a rebuke from the speaker for straying from parliamentary proceedure. Hamburger charged that Webster got his information from James Bailey, "his temporary friend", and that Webster made such "cowardly attacks" only because he was immune while speaking on the floor of the house. [96]

Webster replied that he would make the same charges at any public gathering he could. Unfortunately, no hearings were ever held to look into the charges made by Webster, who may have believed that somehow the Essex water bill was a threat to oust him from his Essex county district.

The Essex county water study bill was defeated, as was a sinister bill to alter the Metropolitan District Commission to eliminate James Bailey from it. The General Court session of 1922 was prorogued on June 13, 161 days after opening. The Swift and Ware Valleys would have to wait at least another year to learn their fates.

Chapter Three   Another Investigation

The water supply issue did not go away during the summer of 1922 -- it was in the minds of many in the western part of the state. Surely it was still on X.H. Goodnough's mind, for he had lost his first battle in attempting to implement his water schemes.

One group that was very impressed by the whole water scheme was the New England Waterworks Association, a society of engineers centered in Boston. Goodnough presented a detailed rehash of the Joint Board report to the group at its spring meeting. Dr. Kelley, J. Waldo Smith, and Professor Whipple of the Joint Board were all present to applaud Goodnough. The presiding officer of the meeting had the naivete to ask of any mill owners or residents of the affected areas were present to comment on the plan. [97]

Even though the water projects cast a shadow over the towns in the Ware and Swift valleys, life still went on much as it did before. Greenwich held a lively Fourth of July celebration, attracting over 2,000 people to the event. Among the featured speakers there were Roland Sawyer and George Webster.

Sawyer extolled the land and people of Western Massachusetts, noting many famous people native to the region. He could not avoid mentioning the water situation; "And now comes a group of men from Boston, who have an insatiable ambition to invade this land; to destroy four of our modern towns, damage six others, and infringe upon the sacred rights of the whole".

After heaping praise upon the residents of the area, he concluded with a dig at Goodnough;

> No man, however entrenched he may be in a powerful department in the State House, can play the Hun in Massachusett and get away with it. No man ... however powerful ... (can) work even greater destruction in the Connecticut Valley than was done by the swarms of the Kaiser's army in Belgium. On this Fourth of July, we reaffirm the right for which the settlers labored amd their fathers died - - the rights to life, liberty, and the pursuit of happiness.

George Webster expressed his pleasure at having an opportunity to visit Greenwich and the other menaced towns. He praised Sawyer, Sen. Griswold, and other local leaders against the water projects. Webster told how he had become interested in the project, through study of the Health Department reports, which he classed as "propaganda". After debunking some of Goodnough's facts and figures, Webster concluded his remarks by calling the engineer "the whole thing in the State Board of Health, and had been its directing force for years. The entire blame for the whole campaign of misleading propaganda must be placed on Goodnough, who distorted facts and figures and perverted reports of his department to meet his own selfish ends". [98]

Prescott held its centennial celebration a few weeks later. Although no politicians were present, a couple of the speakers critized the water plans. One of them labeled the whole question "political". [99]

1923 brought a new session of the General Court, one which would reconsider the water supply plans. The Springfield Chamber of Commerce, with support from the Holyoke and Hartford branches, expressed its reservations about the Swift-Ware diversions. The Chamber was concerned both with the effects the diversions might have on industries, and on the possibility that navigation on the Connecticut River might be extended to Springfield.

Goodnough's water bill was filed with the General Court on January 9th, in the Senate. The next day, Rep. Sawyer filed a bill in the House to authorize the governor to hire an independent engineer to weigh the merits of Bailey's and Goodnough's proposals.

A few days later, the water supply committee heard testimony from all parties on the matter. Both Sawyer and Goodnough ground their usual axes. James Bailey testified in favor of his plan, and he was supported by other speakers. People from Springfield, Chicopee, Holyoke, and the Ware River towns also opposed Goodnough's plan.

On the last day of February, the water supply committee reported "leave to withdraw" on Sawyer's bill. The full house only tabled it to await further developments. On March 22, Senator Christian Nelson of Worcester filed an order in the Senate to have the Joint Committee on Water Supply sit during the recess of the current session. The committee would continue the investigation of the water supply needs and resources of the commonwealth. The committee would be authorized to retain "disinterested and competent engineer(s), qualified as expert(s) on water supply matters". The engineer was also to have had "no connection in any way with any former investigation of the subject under inquiry". 100

Nelson's bill was referred to the Joint Rules Committee of the General Court. This committee held a hearing on April 10, at which Senator Nelson testified that most of the members of his water supply committee were new to the topic and needed extra time to familiarize themselves with it. Senator Wells of Boston, also a member of Nelson's committee, thought that more study should go into the matter, as the engineering features of the Joint Report "were practically the work of one man".

Dr. Glazier of Hudson, another water committee member, thought that it would be a waste to retain another engineer. He felt that the conclusions, not the data of the report were questionable.

Interested parties from parts of the Connecticut Valley recorded themselves in favor of the bill. Some noted that this was all they wanted - - a fair study of the matter by a disinterested party. 101

The Joint Rules Committee sent its recommendation of the bill to the two branches. The Senate passed it that day, under a suspension of the rules. The House followed suit a bit later.

On May 23, the Joint Committee on Water Supply met to arrange for its activities during the coming months. It was agreed to wait for the legislature to appropriate money for an engineer, and the governor's appointment of one. Unfortunately, the General Court never appropriated the money, so the committee had no engineer of it's own. This was to handicap it severely as the legislators could not be expected to interpret the engineering data on their own.

The committee made a trip to the Merrimac River in July, as this stream was suggested as a source for Metropolitan Water. During the last week of the month, the committee, accompanied by X.H. Goodnough, traveled to Western Massachusetts. Stopping at the Wachusett Reservoir, which was a few feet below its maximum level, Goodnough observed, "Oh, this is now down so low, it will never fill again". [102] The Committee made a trip to view Quinapoxet Pond, a suggested supply for Worcester, then went to Coldbrook, to see the Ware River.

The next day, the committee went to the Swift River Valley, it viewed the site of the proposed dam at the Enfield-Ware line. Goodnough gave the committee the particulars of the project, stressing the $60,000,000 cost estimate. This started an argument, as Sawyer and others on the committee disagreed with Goodnough's figures. They cited estimates by M.D.C. engineer Allardice, which put the cost of the project much higher than Goodnough's estimates.

At one that afternoon, the committee held a hearing in the Swift River Hotel in Enfield. Many valley residents, as well as a contingent from Worcester attended. Senator Nelson and Rep. Sawyer both assured the audience that the committee was earnestly seeking a solution to the problem, and would try to do justice to all concerned.

The Worcester representatives stated an urgent need of their city for water, and their feeling that Boston wanted to encroach upon their potential sources. Selectmen from the Swift towns emphasized the point that they couldn't stand the suspense of no definite decision to date. As Selectmen Felton of Enfield stated, "It has stopped an impetus for development, destroyed all enterprise, driven away our youth. We want to see this legislation passed or dropped forever". [103]

Goodnough explained his plan again, emphasizing that the time needed for construction would equal the limit of Boston's present water supplies.

The next day the committee ended its jaunt in Athol, where it heard opposition to plans to use the Millers River to help fill the proposed Swift River reservoir.

The committee also traveled to view the large reservoir being built for the city of Providence, Rhode Island. This project was under the supervision of Frank E. Winsor, of whom we shall hear much more later. New York City water projects in the Catskill Mountains were also visited by the committee.

When the 1924 session began, the committee did not feel prepared to submit a report to the General Court. It also wanted to have the services of a disinterested engineer, as it had heard from Goodnough, Bailey, Dr. Walcott, and J.W. Smith, all of whom represented either of the two major alternatives. An extension of a few months, and an appropriation of $10,000 was requested by the committee.

While this was going on, the Selectmen of Ware were trying to raise a fund from their counterparts in other towns. The money was to be used to defend the interests of the towns along the Ware and Swift Rivers. The scheme did not come to pass as each of the towns involved was more concerned with defending it's own interests. Prescott, out of poverty or resignation on the matter, would never formally hire anyone to represent it in the water fight. 104

The Senate Rules Committee met at the end of January to consider the requests of the Joint Water Supply Committee. There was a wrangle over the proposal to hire an engineer, and whether one would be of any use for only a couple of months. Senator Haigis of Greenfield thought the request for an engineer was a waste of time, as the matter had "been hanging fire long enough".

Senators Nelson and McLane of the water committee countered that the engineer was necessary, and that one could be useful for two months. They cited the views of Governor Cox in support of their statements. A letter from the Boston Chamber of Commerce was produced to oppose the engineer proposal. It stated that it was just a delay to hold up final action for another year. 105

The hearing did not produce much action, and it wasn't until early March that the Senate approved the request to fund an engineer. The house would not pass the budget item, however, so the water committee met again at the end of March. It voted to submit a report to the General Court on the first of May, although the members could not agree on one conclusion or recommendation.

As the committee spent the month of April attempting to come up with a report other big projects in water supply were attracting the attention of Western Massachusetts. What was billed as the biggest earth dam in the world was completed by the New England Power Company in Whitingham, Vt. The 200 foot long dam held back a lake ten miles long, containing 38 billion gallons of water. The whole project cost $10,000,000, including the taking of several farms and homesteads. Projects like these made Western Massachusetts residents speculate that the Swift-Ware schemes were partly for power generation. 106

The Scituate Reservoir, ten miles west of Providence, Rhode Island, was also noted as a parallel to the Swift-Ware projects. In this scheme, a half-dozen villages were being flooded to create a 36 billion gallon reservoir for Providence. It was noted that protests of those living in the affected villages had no influence over the sway of the capital city of the state. 107

The special report of the Joint Standing Committee on Water Supply was issued early in May. Actually, it consisted of three different reports, totaling 22 pages in all. Six members from the House filed the "majority" report. Senators Wells and McLane, and Rep. Pike of Springfield filed a minority report. Roland Sawyer filed his own minority report.

Senator Nelson of Worcester did not sign any of these reports, even though as chairman of the committee and a representative of a water seeking city, he was vitally concerned with the issue. Some officials of the city urged him to support whatever plan would allow Worcester to utilize Ware River water. Nelson replied only that he "reserved his rights" to support whatever proposal might turn out most agreeable to him. 108

A special commission to carry out Goodnough's proposals was the recommendation of the "majority" report. They were apparently sold the idea that the Wachusett Reservoir was built to be "a part only of a larger system" by Goodnough. 109 Proposed legislation at the end of this report moved for immediate action to begin the Ware River part of the project. One member of the three man supervisory board was to be from Boston, one from Worcester, and the other from the Metropolitan District. Another bill at the end of the report provided for an investigation of water for Lawrence and Methuen.

The three man minority report appeared to be motivated by two main considerations. One was the lack of a disinterested engineer for the committee, and the other was the feeling that current Metropolitan water supplies weren't down to the danger point. Two of the rhree men had special reasons for signing such a report. Chester Pike was from Springfield, and represented mill interests who were concerned about reduced water flow to their mills should the Ware-Swift projects go through. Pike added a proposal at the end of the report to allow for the inclusion of a Western Massachusetts member on any board dealing with the water question. Senator Wells was a law partner of James Bailey, the dissenting member of the Joint Board.

A commission to study the water problem, with the aid of its own engineer, was the aim of the three man minority report. They recommended that that the report be filed with the General Court during the 1925 session.

Roland Sawyer's solo dissenting report was longer than the other minority report. He gave five reasons why he couldn't support the majority of the committee. Sawyer felt that there was no emergency forcing action at the time, and that the end of the session was a poor time to rush through such important legislation. He agreed with the minority report in asking for more investigation of the matter, and the retention of a disinterested engineer. His final reason was that "the majority ... have been led away .. into a byway of political intrigue, and that the action they recommend comcommits the state to take sides in the difference between certain public officials. 110

In clarifying his position, Sawyer noted that the Wachusett Reservoir was, at the time, overflowing. He also pointed out the large projects sponsored by Providence and New York City. These, he felt, were part of a trend: "an attractive proposition for engineers, contractors, and politicians. It is soft money from the taxpayers pockets". Sawyer contrasted the proposal to the scandals going on in the Republican administration in Washington at the time.

Sawyer tossed another jab at Goodnough by pointing out that the committee was supposed to have a disinterested engineer, and that "the only engineer we had was Mr. Goodnough, who accompanied us on all trips, and carefully explained the beauties of his plan at every opportunity". 111

Lavish praise for James Bailey and his plan covered the rest of Sawyer's report. He then advocated that a commission (including Bailey) with its own engineer study the question and report back to the General Court by year's end. His bill also provided for Worcester to be represented on the study commission, but oddly did not go as far as Rep. Pike in asking for a Western Massachusetts member.

A few days after the reports were issued the Selectmen of Ware sponsored a meeting of Ware Valley towns to decide on a course of action. Athol residents joined the various Ware River citizens in attendance. It was felt that the cost of property damages should be stressed at hearings in Boston, as this would prevent hasty action on the part of Goodnough Plan supporters. 112

The House Ways and Means Committee held a hearing on the three different bills in the reports on May 19. A host of representatives of Western Massachusetts interests appeared, as well as a man from the Boston Chamber of Commerce. Members of the water supply committee presented arguments in favor of their reports.

The representative of the Boston Chamber of Commerce supported the minority report, in what was felt to be a turnabout from previous advocation of the Goodnough Plan. This may have been due to fear on the part of some Metropolitan residents that Goodnough's scheme would cost too much.

Ben Hapgood of the Springfield Chamber of Commerce argued that the taking of Ware and Swift water would hurt Springfield's industries. He asserted that the State of Connecticut was concerned about the issue, and would call in the Federal government to protect its interests.

W.H. Brooks of Holyoke thought the contention that the majority report was a "constructive" one should mean a "construction" report. Although conceeding that Goodnough was a good engineer, he thought that the man "had become obsessed with his plan, and desires that it be completed as a monument to him".

Springfield City Solicitor Dearborn contended that the legislature had no right to pass a law affecting an interstate navigable stream (the Connecticut). He also noted that industry, river cleanliness, and agriculture would be affected by the Swift-Ware projects.

Speakers from Chicopee and Ware felt that much potential property damage was not estimated in Goodnough's report. John Schoonmaker also spoke for Ware interests in charging that his part of the state had not had a fair hearing on all proposals to date. Rep. Henry Shattuck, chairman of the ways and means committee, asked Schoonmaker to file a bill including a Western Massachusetts member on a three man commission to study the problem.

The bill was drafted, and Shattuck's committee voted to recommend it to the house. 113

As the General Court headed toward the close of another session, it passed a bill to set up a commission to conduct a new investigation into the water question. This was done during the last week of May. The water committee's attempt to get a construction commission passed failed by a three to one margin.

Henry Shattuck told the house that he thought the new investigation would settle most of the disputes about the issue. He noted that all parts of the state concerned would be represented on the commission, and that it would have sufficient funding to hire it s own disinterested engineer of high standing. The commission would have a year and a half to report, giving it plenty of time to do all of the necessary field work. 114

In supporting Shattuck's bill, Roland Sawyer stated "it is time to stop fooling around with the Goodnough propaganda and give the state a cleancut, unprejudiced investigation such as in this bill. 115

Thus was born the Metropolitan Water Supply Investigating Commission.

Chapter Four   A Third Investigation

Two months after Governor Channing Cox signed the bill authorizing the new commission, he appointed three men to serve on it. In keeping with the mandate to have all concerned parts of the state represented, the men came from three different cities.

Elbert E. Lochridge, an engineer with the Springfield Water system, represented the western part of the state. George Booth, publisher of the Worcester Telegram, represented that city. The chairman of the commission, named by Cox, was Charles R. Gow of Brookline.

Gow, who was to give his name to the plan advocated by the commission, was a 52 year old engineer of some eminence. He had served on many metropolitan area boards and commissions while carrying out extensive private consulting and contracting. Gow authored a book, <u>Fundamental Principles of Economics,</u> in 1922, and would write two more on the subject of human engineering. His nephew Frederick was later an M.D.W.S.C. engineer.

While the members of the commission were not known as innovators, they represented as good a choice as could have been made under the circumstances. With two engineers on the board, it would be able to choose a competent chief engineer, and be able to defend his proposals intelligently.

The Metropolitan Water Supply Investigating Commission held its organizing meeting on September 3, 1924. It hired an prominent New York consultant, Allen Hazen, to be its chief engineer. Hazen, born in Vermont in 1869, was an M.I.T. graduate who had been a colleague of Goodnough in the

Mass. Board of Health in the early 1890s. He wrote three books on different aspects of water supply engineering.

Hazen and his team spent the last months of 1924 reviewing the historic trends of the metropolitan water supply, including previous reports on the subject. The team then proceeded to investigate other possibilities for supplies besides the Bailey and Goodnough plans. Included among the alternatives were the Ipswich and Assabet Rivers, both in eastern Massachusetts. Field parties did mapping and test bores at possible reservoir sites on these rivers, and on the Upper Ware River.

After spending $97,000 of its $100,000 appropriation, the Commission issued a report on the deadline date, December 1, 1925. The 177 page report became House Document No. 900 for the 1926 session. One third of the report was taken up by the recommendations of the commission.

The Swift River was acknowledged by the report to be the cheapest supply ultimately, but it was felt that the excessive cost for it need not be born at that time. The report also noted that Swift water would have to be pumped to supply Worcester. Use of the Ware River by both the Metropolitan District and Worcester would solve the problems of the latter and share the cost. Water from the Assabet River could be diverted into the aqueduct from Wachusett Reservoir to augment the Metropolitan supply. A later reservoir on the Ipswich River northeast of Boston would add enough water to supply the Metropolitan system until 1955. [116]

Reaction to the report was mixed. The Metropolitan District Commission seemed in favor of the report (labeled the Gow Plan), as it would keep costs down. Ware River interests were at first sympathetic to the plan, as the taking of Ware water was seen as "inevitable" under all plans. A compensating reservoir near the water supply one on the Ware was planned to keep the mill interests mollified. After studying the report, the organization of selectmen from the Ware River towns and Rep. Sawyer thought that too much water would be taken from the Ware. [117]

Worcester took a similar view. Initially, the plan was seen as a viable way of supplying the city with badly needed water. Then the expense of sharing the cost of the Ware development, as opposed to using "rightful" sources in the Wachusett Watershed drove Worcester into the opposition camp. [118]

The towns in the Assabet and Ipswich River valleys were opposed to the Gow plan from the beginning. Both areas felt that the water available in these streams would prove adequate only for the immediate vicinity. [119]

The Joint Legislative Committee on Water Supply scheduled its first hearings on the report in early February, 1926. Meanwhile, most of the municipalities involved were hiring attorneys to defend their interests. The Ware valley towns appointed an engineer from Lowell to study all reports for them. Even Prescott felt the need to appropriate $100 of the town's meager funds for legal expenses concerning "the matter of the metropolitan water taking." [120]

The first hearing was postponed due to bad weather. George Booth of the Special Commission was thanked for attending anyway. He had spoken in defense of his group's plan at the Boston Chamber of Commerce the day before.

A week later, on the 17th of February, hearings finally got underway. A host of speakers came in support of the Gow plan, including the Commission members and Allen Hazen. Only Rep. Sawyer and X.H. Goodnough opposed the plan. Davis Kenniston, of the Metropolitan District Commission, argued that the Gow plan represented a "middle course". and would meet all needs for water for a couple of decades. Worcester representatives asked that something be done as soon as possible to ease their water shortage. Roland Sawyer picked at parts of the plan affecting the Ware River. Charles Gow defended the plan as the best way to develop water sources with the least interference to private interests. 121

Another hearing took place the next day. Essex County towns presented a unified front against the use of the Ipswich River by anyone but themselves. They argued against the Gow plan's contention that much land in the Ipswich watershed should be purchased immediately and held for future water supply use. Essex people also thought that their region was not growing as fast as the report implied, hence their property was not going to see a huge increase in value. Finally, they pointed out that no hearing was held by the Commission in their region, as required under the enabling act. 122

During the hearing, Hazen was asked about the accuracy of the information presented in the Joint Board report. He stated that while he might disagree with some of the conclusions of the report, he could not find anything wrong with the contents.

Roland Sawyer quizzed the Worcester interests on their support for the Ware River supplies when previously they had favored using the Quinnepoxet River. He drew an admission out of the Worcester City engineer that the Ware development would be cheaper for the city if shared with the Metropolitan District. 123

Former Speaker of the House B. Loring Young appeared for the town of Framingham, and asked that no more encroachment on that town be made by adding to watershed lands there on the Sudbury system.

The next round of hearings took place on February 22nd. This time it was the turn of the Ware River interests to assail the Gow plan. Former Attorney General Herbert Parker, as special counsel for five towns in the Ware Valley, claimed that the report discrimminated in favor of Worcester. He satirically pointed out that the "dark brown characteristics of the Ware River, so strikingly pointed out by Worcester's own commission in . . . 1920, takes on a golden hue now that the means have been found to force the Metropolitan District to assume the larger . . . cost of meeting the city's needs. 124

Parker critized the lack of provision for eminent domain taking damages

PERSONALITIES

Rep. Roland D. Sawyer

Rep. George P. Webster

*L. to R.: X.H. Goodnough, Unknown Engineer, Karl Kennison,
Frank E. Winsor, Davis Kenniston
(Courtesy S.R.V.H.S.)*

DANA COMMON

NORTH DANA

VIEW OF ENFIELD

SMITH'S VILLAGE
Post Office and Mill

GREENWICH VILLAGE

GREENWICH
Congregational Church

*NORTH PRESCOTT
in the Winter*

*MILLINGTON
Post Office and Store*

COLDBROOK SPRINGS
Baptist Church

WINSOR DAM
Under Construction

AQUEDUCT
Intake Shaft Before Flooding

WINSOR DAM
Looking West

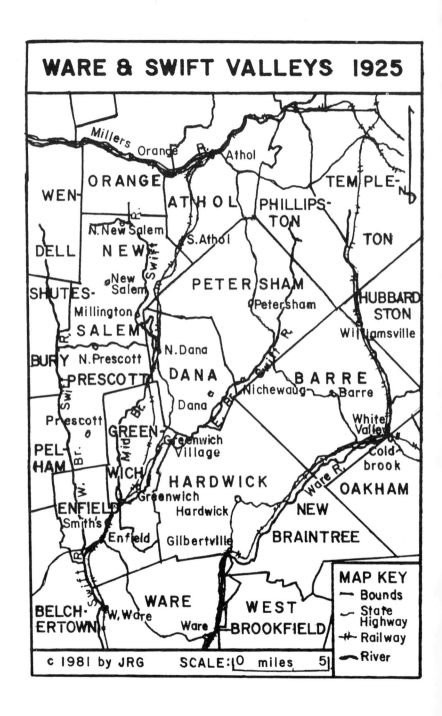

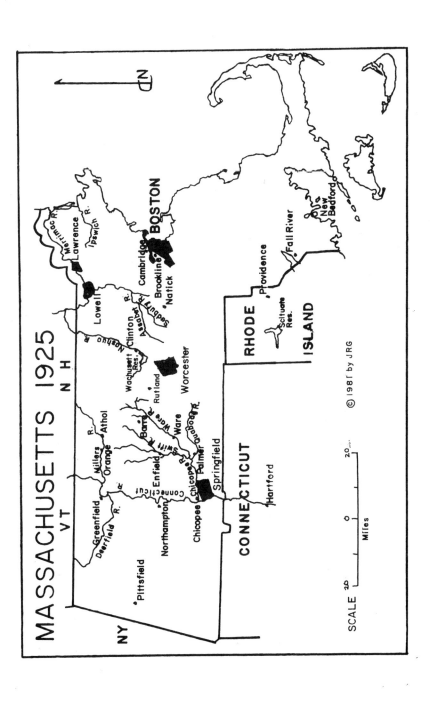

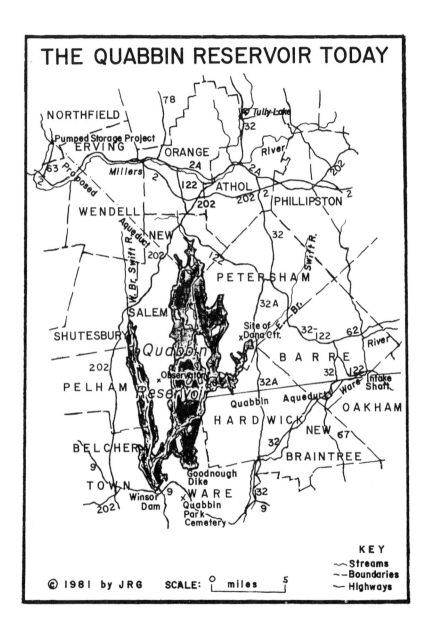

in the Gow plan, and felt that the compensating reservoir was inadequate. Attorney George Storrs of Ware took up Parker's theme, and emphasized the potential damages to the numerous mills along the Ware. He also produced figures to show that the compensating reservoir would be too small. John S~~cho~~ Schoonmaker of Ware also defended the mill interests, and added that the compensating reservoir was not even a mandatory feature of the legislation.

A speaker from Rutland stated opposition to the plan only in that it did not provide adequate compensation for damages in that town. [125]

Two days later, when hearings resumed, over 150 residents of Framingham packed the committee room, prompting Chairman Christian Nelson of Worcester to remark, "Is the whole town here?" Framingham speakers scored the Gow plan proposal to take filtered water from the Sudbury River to augment the Metropolitan supply. They contended that enough water left their town, and that local industries would suffer harm from the decreased flow in the river. [126]

Dr. H.B. Arnold of the Boston Chamber of Commerce voiced the support of his organization for the Gow plan. He felt that the immediate needs of the city of Worcester and the Metropolitan district would be best served by adoption of the plan. In answering an inquiry about the proposed sharing of costs, he said his group accepted the Metropolitan burden of eight ninths of the cost as being fair.

Speakers from Barre, Hardwick, and Oakham scored various aspects of the Ware diversion proposals, especially the compensating reservoir. The hearings then adjourned for another week.

Charles Gow spent over two and a half hours defending his report when hearings resumed on March 2nd. He went over the objections to each of the four main parts, asserting that it had been designed to harm private interests and industries as little as possible. Gow also noted that it cannot be expected that any locality would willingly give up some of its water to another without opposition. Finally, he cited the relatively low cost of damages incurred by the construction of the Wachusett Reservoir as reasoning to rebut the high estimates made by Ware River interests. [127]

Gow was followed by a stream of officials from the Assabet River region. They felt that Marlboro had gained little benefit from the 800 acres already lost to the Metropolitan District, and that further incursions would choke off the town's growth. It was also noted that the water supply of the town was running low, so that the river might soon be needed by Marlboro itself.

At this point, Rep. Haskins of the joint committee asked "What in God's name can we do to get water for the Metropolitan district?" This referred to the opposition from all regions concerned to the Gow plan. [128]

A reply by Rep. Mullen, was that House bill 900 was a weak substitute for the Joint Board report. Former Rep. Charles Gilmore of Melrose then made a comparison between the two reports, in favor of the Joint Board one. He pointed out that the Gow plan gave filtered water, less water, and a greater ultimate expense than the other plan. To keep Worcester satisfied, he proposed pumping water from Wachusett Reservoir into the city, with the power needed to be supplied by the dam at the proposed Swift River reservoir. [129]

Roland Sawyer issued a rebuttal to Gow's speech the next day, entitling it "The Cat out of the Bag". He charged that the interests of Worcester were utmost in the minds of the Special Commission, and the the Ware River was included chiefly to appease that city. He went on to accuse Gow of willingness to give up either the Assabet, Ipswich, or both to secure the Ware and the Swift, and appease the Metropolitan District in the bargain. Sawyer saw Booth, the Worcester member of the Commission, as calling the shots, and asked why Lochridge of Springfield sat back and did nothing. [130]

On the 5th, the Senate Ways and Means committee, which had the water bill for the upper house, voted to send it to the floor. While it did not view the bill as being totally acceptable, they felt that any amendments to it should be made on the floor of the Senate.

One committee member, A.B. Rice of Boston, stated he would amend the bill with a proposal by J.R. Freemen, a consulting engineer for Boston. That proposal would alter the Ware River part of the project. Senator Gaspar Bacon and others intended to substitute the Goodnough plan for the whole Gow bill. The House Ways and Means Committee would take its time in studying the measure before sending it to the house floor. [131]

A few days later, the joint committee met again. They anticipated a long statement by Boston Mayor Nichols opposing the Ware project, but he notified them he would submit his opinions in writing. He later made a statement to that effect, but qualified it by his belief that it would have no influence.

W.H. Brooks of Springfield, representing industries, defended the Gow plan, but asked that the compensating reservoir on the Ware be made mandatory in the legislation.

Allen Hazen, with Davis Kenniston of the Metropolitan District Commission, spent the rest ot the hearing rebutting arguments against the Gow plan.[132]

After this hearing, there was much sentiment for an official tour of all the areas involved in the proposed water projects. Although the committee was not in total agreement on the need for a trip, a request was filed with the leadership of both houses of the General Court to issue travel orders. Approval for the trip came at the end of March.

The committee seemed more interested in the trip than anything else as March ended. At another hearing, representatives of the town of Ashland were allowed to ask for safeguards for their town in the legislation relating to the Sudbury River. However, when former Attorney General Parker and a colleague wanted to rebut the Hazen-Kenniston statements of a week earlier, the committee told them to do it in writing. [133]

Senator Christian Nelson, joint committee chairman, set up the water tour to begin on April 1st. The Metropolitan Affairs committee was invited to come along. Among other state officials on the tour were W.E. Foss and Davis Kenniston of the Metropolitan District Commission, Karl Kennison of the Health Department, and the three members of the special commission. Boston water department representatives were also on hand.

The group set out by automobile, and stopped first in Framingham. Then they proceeded to Worcester, whose underfilled reservoirs they viewed. A swing up to the Ware River in Barre was completed before returning to Worcester to spend the night. Worcester showed its appreciation for the visit by putting on a banquet for the visitors, hosted by the mayor. [134]

The next day, the group went from Worcester to Enfield, to view the Swift River Valley. X.H. Goodnough's subaltern, Karl Kennison, explained the 1922 proposal at the site where it would be carried out. After lunching in Enfield, the group went back to Boston, swinging through the Ware River valley and by Wachusett Reservoir on their way back. [135]

Senator Nelson called another meeting of the joint committee for April 7th. In an executive session, the committee dealt with the request of Connecticut Governor John Trumbull to present testimony to the committee. A representative of Trumbull had been following the course of the hearings, and had communicated to the governor that this was the time to present his case.

Connecticut was concerned about possible effects that diversions from tributaries of the Connecticut River might have on their state. The governor wanted to have his opinions on record before any legislation was passed, so that court action could be taken against Massachusetts if necessary. [136]

Before the scheduled meeting began, committee members disagreed on how to conduct further business. Several members, led by Rep. Gilman of Boston, wanted the Governor, Alvan Fuller, to mediate the issue. Gilman informed the governor that "Worcester is attempting to get a water supply at the expense of the Metropolitan District.". [137]

Senator Nelson hotly disputed these remarks, revealing that the committee was openly splitting into factions. This resulted in a conference of committee members and engineers in the Governor's office during the afternoon. X.H. Goodnough and his assistant Karl Kennison pushed the former's proposals. Some of those present protested this, causing the governor to set up a series of meetings for the next day. Fuller promised to hear representatives of all views, then decide which proposal he felt most comfortable with. Many on the committee felt that the Governor's feelings should be known, since he would have to sign whatever bill was passed. [138]

Actually, the request to have the Governor make a decision after only two days of consultations was probably more of a buck - passing operation than anything else. The Governor could be conveniently blamed by the losing side if his decision went against them. It was also a good ploy by supporters of Goodnough to get his plan back into consideration by influencing one man. Fuller was not a strong personality, as seen by his vacillations on the Sacco and and Vanzetti case. [139]

The next day, Fuller met with a large group for over six hours to try to hammer out a compromise. The group consisted of Goodnough, Kennison, Gow, Kenniston and Foss of the M.D.C., J.Waldo Smith, Rep. Davis of the House Metropolitan Affairs committee, and Senators Nelson and Warren of the Senate Water Supply Committee.

Initially Fuller wanted to meet with all parties separately, but it ended up being a conference. Kenniston argued for a modified version of the Gow plan, as he had all along. Goodnough and Kennison pushed the former's plan. Senator Nelson and Gow spoke for the latter's plan. Fuller ended up without committing himself to either side, and asked for a group of the engineers involved to attempt a compromise. [140]

Whether consciously or not, Fuller cast his lot in with Goodnough when he selected the engineers. J. Waldo Smith, Goodnough, and Gow were to form the engineers group, with Davis Kenniston sitting in. Smith and Goodnough had previously been in accord on the latter's plan. Kenniston, who was only to be an advisor, did not fully support Gow's plan. Gow himself had been having second thoughts about the proposal that bore his name, fearing that it might not prove adequate for a long time without the Swift River involved.

That same day, down the hall from the Governor's office, the joint committee held a hearing to listen to Caleb M. Saville, the representative of Connecticut Governor Trumbull. Saville, formerly connected with the Metropolitan Water and Sewerage Board, read a prepared statement about the effects diversion from Connecticut River tributaries would have on this state. He noted that river pollution would concentrate more in the reduced flow, and that the natural scouring effect of the river would be harmed. Saville also pointed out that industrial use of the river and navigation would be affected by reduced flows.

An implied threat of future water diversions was elaborated upon by Saville," we see in this taking of a portion of the Ware River only the beginning of a larger plan outlined in 1895. Where else can the Boston Metropolitan district ultimately go?" He did not wish to sound belligerent, but wanted his state's view heard. A conference of the two states was his suggestion for ironing out any disagreements.

When asked if his state would start legal proceedings against Massachusetts if the water bills passed, Saville replied, "Soveign states are bound to self-preservation and to the protection of their own citizens". [141]

Charles Gow, before meeting with the other engineers in the Governor's group, conferred with George Booth, the Worcester member of the special commission. This apparently gave Gow some backbone in supporting his original plan to benefit Worcester. 142

The next day, the engineers met with Fuller and gave him two options, one of which was a weak alternative to the other. Among the statements in their report were the following:

> The theory upon which the special commission has worked ... involved the conserving of all possible existing and related supplies in order to defer ... the substantial investment required for a long term solution... As a part of such a plan ... the Assabet and Ipswich Rivers were included ... if these are eliminated from consideration the problem presented is materially different....
>
> Under these circumstances, it seems to us that there are only two lines of proceedure which are worthy of consideration ... the complete development of the Ware and Swift River sources in substantial accordance with the plans of the Joint Board....
>
> The alternative proposal would be to take the Ware River water, in substantial accordance with the Special Commission' report ... it might defer the subsequent step of building to the Swift for a period of four to five years at the outside.
>
> In case a satisfactory financial agreement was effected between the city of Worcester and the Metropolitan District which would permit the Metropolitan to retire from this source (the Ware River) ... this alternative would be preferable to Mr. Gow. 143

Obviously, Gow had given in to those in opposition to the Assabet and Ipswich parts of his plan, and in doing so, was committing himself to the Goodnough plan, sooner or later. Upon hearing this, many members of the legislative committee also withdrew support for the Gow plan, feeling that there was no other course to take.

Governor Fuller was satisfied with the report, and expresed the opinion that the five year controversy would come to an end. He did this at a conference in his office, attended by most of those who had met with him a few days earlier. It was agreed that Worcester would get another hearing on this new report before the legislative committee a few days hence.

Reaction to the report was favorable in the Ispwich, Assabet, and Ware Valleys. The former two regions were happy to be dropped from consideration, while Ware valley people still wondered how much water Worcester might want from their river. A spokesman for the Ware interests stated that they "had no particular reason for opposing the Goodnough proposition so far as the Swift River is concerned". 144 The Swift River valley would now be alone in it s fight for survival; its fellow opponents had been divided and conquered effectively.

Chapter Five   More Politics

On April 13th, representatives of Worcester and the Metropolitan District got together to form plans for the divided use of the Ware River. The result, named the Booth plan for the special commission member, was that Worcester would initially pay two-ninths of the cost of the Ware project, and gradually take over the whole thing within twenty years. Presumably by the time Worcester took over the Ware, the Metropolitan would be using the Swift as a water source.

Worcester representatives still favored the original Gow plan with respect to supplying water to their city, and they were supported by the testimony of Allen Hazen. Booth pointed out the city was not a part of the Governor's special report committee of a few days earlier, so that his proposal should receive separate consideration.

That evening, Roland Sawyer issued a statement attacking the Booth plan. He charged that it was still another way for Worcester to get the Metropolitan District to pay for its water. Sawyer viewed this plan as "a little less harsh bargain". Of course, his motive was defending the Ware River from being diverted too much by Worcester or Boston. 145

After intense lobbying, Worcester was all set to get the Joint Committee on Water Supply to vote the Booth plan officially on the 16th. This was dashed at the last minute by Boston's request that it be allowed to testify on the matter. Mayor Nichols, who had declined to speak on the subject earlier, now led the call for Boston to be represented by other than its legislators in the debate. John R. Freemen, the noted Providence, R.I. engineer, was retained by the city to represent its interests.

A delegation from the Ware River towns appealed to the Governor to beware of the Booth plan. They asked him to keep the interests of their region in mind by supporting some other compromise. The Governor gave the group a vacant assurance that he would "have due regard for your position." 146

Although no Swift River people attended the meeting with the Governor, they expressed their opposition to the project through local forums. Rep. Haskins, although not rabidly promoting his constituents' best interests, made known their displeasure to the General Court and the Governor. 147

A week later, on the 26th, the water committee and Metropolitan Affairs committee, sitting jointly, split in two on the water problem. This reflected what would become a split between the House and Senate on the question.

After hearing John R. Freeman propose a variation of the Booth plan, the committee voted on what plan to recommend as a bill. Freeman's proposal lost by a vote of 12 - 7. The Booth plan lost by a closer margin, 11 - 9. The only proposal voted in was the Ware River portion of the Goodnough plan, which won by a vote of 10 - 9. Both Haskins and Sawyer voted for the latter proposal. Since five members of the committee were absent, a vote would be held the next day to make a final decision. 148

The committee's vote was reversed the next day, when all members attended. Voting only on the Booth or Goodnough plans, thirteen supported the former, and eleven the latter. Both sides prepared bills for submission to the Senate Ways and Means Committee. That committee was seen as supporting the Booth plan, as two key members were on its side as members of the water committee.

The Goodnough plan had a better chance in the House, as most of its proponents on the joint committee had been House members. Rep. Sawyer commented that the report of the joint committee had little value, because of the "peculiar conditions" surrounding its origin. He again charged Boston legislators with having sold out to Worcester to the detriment of their own city. 149

The bill for the Booth plan called for a commission consisting of the M.D.C., with one member from Worcester added. This body would supervise the construction of Ware River works for both Worcester and Boston, as in the Gow plan. Worcester was to pay two-ninths of the initial cost, and take over the whole works by 1945.

The other bill filed was to initiate the first part of the Goodnough plan, which would build an aqueduct from the Ware to Wachusett Reservoir. Worcester was to receive water from the Wachusett. The construction commission in this bill was to consist of a Bostonian, one citizen from the rest of the Metropolitan District, and one from some other part of the Commonwealth.150

When the Senate opened its session on the 28th, a few of the Senators on the Ways and Means Committee, led by Walter McLane, tried to have the rules suspended. By this ploy, they tried to have the full Senate accept the Booth plan without any committee hearings. This was voted down by a 22 - 5 margin.

After the vote, there was a row between members of the Ways and Means Committee and chairman McLane. Senator Hartshorn called the committee a joke, and Senator Rice stated it would be a "one-man committee" no more. McLane's suggested hearing date was pushed back a few days at the insistence of the other committee members. 151

The Senate Ways and Means Committee met on April 30th. It heard many objections to the Booth plan from Ware River interests, headed by former Attorney General Parker. Parker protested the way the bill was endorsed by the committee without consulting his clients' interests.

Chairman McLane stated that his committee had nothing to consider except the "financial end" of the bill. Parker argued that the merits of a project should be considered along with the financial aspects.

A large group from the Ipswich and Assabet towns expressed fear that the adoption of the first part of the Gow plan (in the guise of the Booth plan) would lead to later use of their rivers. [152]

A spectator at the hearing charged that Charles Gow was a tool of Charles Innes, the Republican boss of Boston. He asked, "Don't let the boss of Boston be the boss of Massachusetts". This referred to Innes' earlier lobbying for Worcester's favorite plan. Gow, who was present, did not dignify the charge with a reply. [153]

The next hearing, held on May 3rd, saw a number of rebuttals, by familiar speakers, of earlier arguments in favor of the Booth or Goodnough plans. Charles Gow, Davis Kenniston, and Allen Hazen all spoke on the merits of the Booth plan. Gow defended his supposed "desertion" of the plan that bore his name by arguing that he was supporting a plan based upon the work of the special commission.

Allen Hazen, in a statement characterized as being thoughtless, asserted that if there already was a tunnel from the Swift to the Wachusetts, Worcester would find it cheaper to build its own works from the Ware River than to pump from the tunnel. This was seen as hypocritical when compared to the plan Hazen recommended to the special commission. [154]

Schoonmaker and Storrs, representing Ware River interests, argued that the compensating reservoir planned for them was inadequate, as were the damages provided for in the Booth plan bill. Hazen broke in to cite figures proving the compensating reservoir would be adequate. Storrs admitted that it might be, but that no guarantee to this effect was in the bill. He also cited an unnamed Connecticut source reporting that a compensating reservoir for a river there failed.

Counsel for the town of Barre felt that the bill did not adeqately compensate for all real estate the water project would remove from taxation. Ipswich and Beverly, in the Ipswich basin, recorded their towns against the Booth plan.

While members of the House Ways and Means Committee attended the Senate session to avoid duplicating testimony, they postponed any action on the bill for a week. [155]

The next day, several engineers testified about the two plans under consideration. John R. Freeman, representing Boston, restated his unpopular plan. X.H. Goodnough and J. Waldo Smith supported the former's plan. Allen Hazen defended the Booth plan. Robert Weston, a private consulting engineer, displayed examples of Ware River water compared with other water to prove it was acceptable for use as a water supply.

On the 5th, in executive session, the Senate Ways and Means Committee voted unamimously to send the Booth plan bill to the floor. Members reser the right to amend it, feeling that it was not totally acceptable in its present form. 156

The Ways and Means Bill came to the Senate floor on May 7th. Senator Gaspar Bacon of Boston led the fight to have the Goodnough plan subsituted for the Booth plan. Senator Haigis of Greenfield came out for the Goodnough plan, citing his belief that the question of the Swift River ought to be "settled for the sake of the people in that valley".

Senator Rice of Boston wanted to postpone action until Boston could have more say on the matter. A voice vote turned down his request. Senator Nelson then argued in favor of the Booth plan, stating that it would cost Boston less to ger water that way.

Bacon's request to substitute the Goodnough plan passed on three successive votes, ending with a 19 - 8 majority. This sent the matter back to the Ways and Means Committee, as a new bill was involved. George Booth, who witnessed the Senate vote, issued a statement charging that Worcester was unfairly treated by that day's results. 157

The Ways and Means Committee met on May 10th to consider action on the Goodnough plan. The city of Boston voiced support for the plan, as did Worcester, providing their water rights were provided for. Senator Rice's last attempt to push the Freeman plan received no support. A member of the Hampshire County Commissioners urged that his county be protected from damages to be sustained in building a Swift River reservoir. Herbert Parker noted that Ware interests would accept the Goodnough plan with added damage provisions. Davis Kenniston wanted water as soon as possible, suggesting the filtration of Sudbury River water if nothing else could be put through. Somerville opposed giving Worcester any Metropolitan water.

After these statements, Senator McLane declared his view that the Gow plan was "dead", although its damage provisions could be written into a bill for the Goodnough plan. His opinion that the committee should ask to be relieved of the task of writing the bill was seconded by the rest of the committee. 158

A compromise was worked out that same day between representatives of Worcester, Boston, and other parties. It stated that Worcester would get to take the Quinnepoxet River from the Wachusett watershed for its own use. The Metropolitan District would be paid $800,000 for this water, and allow Worcester to have rights to waters in the northern reaches of the Ware watershed. This proposal formed the basis of legislation that was taken up later. 159

On May 13th, three sets of amendments to the pending water bill forced a conference in the Senate President's office. The most controversial topic was the proposal to give Worcester water rights in the Ware watershed. Since no agreement was forthcoming, action was delayed a few days.

The next day, the House announced a plan whereby it would take action to pass the Goodnough plan. It charged the Senate with trying to delay the matter until the 1927 session. The House Ways and Means Committee was prepared to hold hearings and act on the Goodnough plan, thereby forcing the Senate into action. Questions were also raised concerning the composition of any special commission to construct the projects. The main bone of contention there was whether to include members who were not engineers, or from outside of the Metropolitan District. Western Massachusetts interests charged that they would not be properly represented by a metropolitan commission. The Metropolitan District Commission was felt by some to be inadequate to the task of constructing new waterworks, as it was more of a maintenance operation. 160

On Monday the 18th, the Senate debated the water situation for two hours, then ordered the water bill for a third reading. This was, in the words of Senator Bacon, a "compromise bill". The compromise was that Worcester's right to take the Quinnepoxet River was retained, but the cost was raised to $1,250,000 by Senator McLane's amendment. Another amendment, by Senator Haigis, passed by an 11 - 9 margin, and provided more protection for the Swift and Ware valleys. 161 Two days later, the Senate finally passed a wat-water bill and sent it to the House. Senator Haigis' amendmets of the other day were excised, and replaced by a vague amendment of Senator Rice's superficially protecting Ware - Swift landowners. Worcester's votes and supporters were won over against the Haigis proposals by the reduction of the Quinnepoxet price to an even million dollars. Amendments to define what floodwaters of the Ware could be diverted, and disallow Worcester from getting any of them were lost, as was one to put Worcester and Western Massachusetts representatives on the construction commission. 162

Rep. Sawyer lambasted the Senate's actions in a long statement at the end of the day:

> The story of today's events in the Senate is simply a record of the efficiency of a steam roller. The steam roller got in motion over Sunday and first eliminated the Goodnough elements from the building of the proposition, then working westward it yesterday flattened Worcester, and today still working westward it ground all rights of Western Massachusetts into a pulp . . . . the record of swapping and jockeying in the Senate over the past few days is pathetic.

Sawyer went on to suggest that the Senate go back to either a "straight Goodnough plan", or postpone the issue to 1927. In his desire to get the best deal possible for Ware, Sawyer warmly defended Goodnough's proposals, and attacked the M.D.C. for its "rapacious" settlements with landowners in previous dealings. 163

The House accepted the Senate bill for consideration, and referred it to the Ways and Means Committee. At its hearings the next day, the committee heard Davis Kenniston of the M.D.C. assail the Senate bill as being the worst proposal and a waste of money for all sides. He was supported by Roland Sawyer, who also objected to the makeup of the commission proposed, and poor settlement provisions for affected landowners.

Nelson Matthews, a former Mayor of Boston, assailed the bill's provisions for damages for loss of business or employment. "No act ... can be cited as precedent for such a provision", he added. He and the other present forgot about the Wachusett Reservoir bill, and how thousands of dollars were paid out for just those reasons.

A few other speakers came out mostly against parts of the bill. Then representatives of the Swift River valley were heard. They wanted some Western Massachusetts representatives in the construction commission and adequate compensation for their forced removal. F.E. Parsons of Enfield noted that no area banks would lend money for real estate in the valley. J.H. Johnson of Dana said "We have been in the valley of doubt long enough". 164

Worcester Solicitor Mellish stated that his city wanted a separate water supply, not wanting to be hooked up with the M.D.C. He quoted Goodnough on the possibilities of using North Ware water for the city. A selectmen from Rutland supported that possibility. 165

The results of the hearing became apparent on the 24th, when the report of the Ways and Means Committee was submitted to the House. Many changes were made in the Senate bill. The proposed building commission was to be chaired by the M.D.C. chairman (Kenniston), but with outside members. The employment loss provision was omitted, but towns along the Ware River were given the right to obtain water from the new works. A provision was inserted to provide for the Sudbury water filtration project, to secure an immediate added supply for the Metropolitan District. Finally, the employee status of the construction workers was clarified, and a definite limit for the taking of the Ware River was set.

After a brief debate, the House ordered the bill to a third reading the next day. Rep. Pike of Springfield tried to substitute a plan to filter Sudbury waters alone for the bill, but this was defeated. Henry Shattuck of Boston read a long speech summarizing the bill and its history as an introduction. Rep. Twohig of South Boston mocked Shattuck by stating that ending prohibition and allowing free beer would ease the water problem, but he ended up moving the previous question. Senator Warren's handout of a letter from John Freeman (containing the latter's proposals) was mostly ignored. 166

On May 26th, the Senate passed the House bill, with only one small amendment. At the same time, an amendment by Rep. Haskins was passed, limiting the diversion of the Ware to the winter and early spring months. Two days later Governor Fuller signed the so-called Ware River Act into law. He gave the pen used in the signing to Leslie Haskins, which was much appreciated by the latter. It didn't help Haskins' political career, however, as he was defeated for re-election that autumn. 167

Chapter Six   The Swift River is Taken

Chapter 375 of the Acts of 1926 (the Ware River Act) contained an emergency preamble; thus it went into effect immediately upon being signed by the Governor. However, the administrative machinery it provided for would take a few months to set up. It is important to note the agency the act created, and who staffed it, as this body was to be of paramount importance to the Ware and Swift valley residents for the next dozen years.

In the first section of the law, the Metropolitan District Water Supply Commission (hereinafter referred to as the M.D.W.S.C.) was to be strictly a construction commission, its works to be built for the eventual use of the Metropolitan District. The chairman of the M.D.C. was to be the chairman of the new board, with two associate commissioners, all serving five year terms and appointed by the Governor. The chief engineer for the M.D.W.S.C. was to be appointed by that body.

To speed its work, the commission was allowed to purchase land in the Swift River Valley, but it had to draft enabling legislation for the Swift project by 1927. Bond issues for $15,000,000 were authorized to fund the Ware project.

Davis Kenniston, who had chaired the M.D.C. for a year and a half, was duly appointed chairman of the M.D.W.S.C. Charles N. Davenport, of Boston, and Joseph W. Soliday, of Dedham, were appointed as the associate commissioners, all on July 28, 1926. Two days later, the Commission formally organized, meeting in the M.D.C. headquarters. The three commissioners proceeded to set up the administrative machinery for the projects, including the appointment of various engineers. The choices for consulting engineers showed who would have great influence on the project. X.H. Goodnough and J.W. Smith were the regular consulants, while C.T. Main advised on riparian matters. Thus, both the M.D.C. and the Health Department had influence over the course of work. R.Nelson Molt was named secretary for the M.D.W.S.C.

After deliberating for a month, the commissioners appointed Frank E. Winsor as chief engineer. This man would be the one figure most associated with Quabbin Reservoir in the minds of the general public, so a brief biography is in order here.

Winsor was born in Providence, Rhode Island, on November 16, 1870. He received a Bachelor's Degree in Philosophy from Brown University in his home city in 1891, and a degree in civil engineering a year later. After marrying a hometown girl (they would have three children), he worked on the construction of the Metropolitan sewer system in Boston. In 1895, he became a design and construction engineer on the Wachusett Reservoir job, staying until 1903.

That year, Winsor was placed in charge of a field survey of 100 miles of

New York City reservoirs. Near the end of the year, he returned to the Boston area to become deputy chief engineer for the Charles River Basin Commission. From 1906 to 1915, Winsor was back in New York, supervising the construction of the Kenisco and Hillview Reservoirs, and part of the Catskill Aqueduct. The construction of the reservoirs was a $35,000,000 undertaking, with J. Waldo Smith as one of the engineers.

1915 saw Winsor move back to his native Provicence, as chief engineer of the Scituate Reservoir project. This work was just coming to a successful conclusion when he was hired by the M.D.W.S.C. His salary in the new position was $13,500 a year, a hefty sum for the times.

To serve under him, Winsor brought in a number of pet engineers from the Providence job, and Goodnough got many of his friends and underlings positions. N. Leroy Hammond, Karl Kennison, and Arthur Weston were but three of Goodnough's men on the project. A minor engineer, F.D. Farley, had been associated with the Ashokan project in New York twenty years earlier. 168

Field offices were set up by the commission in Enfield, Hardwick, and Holden to begin work on land and engineering surveys. The first offer to sell land was received in October, 1926, and the first settlement made in December. A firm was hired to do an aerial survey of the Swift and Ware Valleys to develop data for land aquisition. The draft of legislation for the Swift Reservoir was also in preparation.

While the M.D.W.S.C. was putting itself together, the citizens of the river valleys were contemplating their fate in a rather leisurely manner in the summer of 1926. Much disappointment was expressed at the lack of local representation on the water commission, but no formal protest was ever lodged with the Governor. Leslie Haskins was running for reelection as a state representative, with an opponent from Athol. Although he received endorsements from Swift Valley residents outside of his district, he lost the primary election. Roland Sawyer won reelection handily, but expressed disappointment that Enfield, Greenwich, and Prescott did not give him much support. 169

In the late fall of 1926, the selectmen of the Swift valley towns formed an association of their boards to promote their interests when the legislation on the valley came up. They recommended various men to serve as real estate appraisers for the M.D.W.S.C. Various plans for creating a new town or towns out of valley territory were considered by the group, as was the possible relocation of the rail line that ran through the valley. This group soon evolved into the Swift River Valley Protective Association, with the participation of many valley residents. Leslie Haskins chaired the organization. 170

Just before the 1927 session of the General Court was to commence, the association prepared a list of demands in regard to any legislation relating to the valley. Tentative approval of the list was obtained from the M.D.W.S.C. The association wanted a deadline of March, 1928 for all property taking involved, protection for employees of valley businesses who might lose work time, and that proper arrangements be made for the removal and reburial of all bodies currently interred in valley cemeteries. 171

A meeting of the association members and the M.D.W.S.C. in early January, 1927 left Haskins and other members discouraged. The commission did not choose to obligate itself to buying up all the required property by March, 1928. It did not support the idea of compensating business and professional men for the loss of their livelihoods; even if they lived within the bounds of the proposed reservoir. 172

A 100 page bill was filed by the M.D.W.S.C. in the General Court in the second week of January, 1927. It set up the system by which land was to be purchased for the reservoir, but no deadline was set. Blue collar workers were to be safeguarded from employment loss for up to six months, using the Wachusett Reservoir method as a precedent. Boards of referees would be set up to hear disputed cases involving takings of land and water rights. Roads, buildings, cemeteries, and other public property could be removed or relocated.

Upon receiving copies of the bill, Haskins and other association members began to prepare amendments to it. One that they felt strongly about was to have the M.D.W.S.C. compensate valley residents for damage done "during the period that the state has talked and done nothing about taking the valley". Although realistically feeling that such an amendment had little chance of passage, any retroactive time period would be better than compensation paid only at the time of any taking. 173

In the next couple of weeks, while engineers were doing test drillings on the Ware-Enfield line, Haskins and his cohorts succeeded in convincing members of the legislative water supply committee to hold a hearing in Enfield. The Association met a few days before the hearing to plan strategy, which included having Attorney Storrs of Ware attend as their counsel. 174

The hearing was held on Thursday, February 3rd. Several hundred people packed the town hall in Enfield to note the proceedings. Besides the legislative committee (which included Reps. Worrell of Athol and Sawyer of Ware) all three of the M.D.W.S.C. commissioners were present. Attorney Storrs made a sentimental appeal to have nine amendments added to the bill to protect the valley's rights. Storrs said "We want you to take this view of it - - as a bargain and a sale - - you would give and take and arrive at a conclusion". He spoke of the "clouds hanging over the community" for over 30 years, which put a damper on local growth.

Charles Felton, Enfield Selectman, asked for a quick end to the project, stating that "Five years is a long period at my time of life". Dr. Segur of the same board seconded this, and asked for retroactive damages for the valley residents.

Charles Cooke and Attorney William Duncan of Athol were joined by J.H. Johnson of Dana in asking for the relocation of the B & A Branch railroad.

They felt that the loss of it would seriously affect the industries of areas on the northern fringe of the reservoir. Concern about taking the flood waters of the Millers River was also expressed by the Athol men. Other speakers were heard, including a group of women, before the committee adjourned to spend the night at the Swift River Hotel.

The next day, the legislators drove up the valley to Athol, where they lunched at the Athol House on Rep. Worrell's invitation. The importance of an undiverted Millers River and the B & A Railroad to the town were stressed. The committee declined to state any opinions on what they heard, but set another hearing date for the next Tuesday in Boston. [175]

Before the next hearing was held, a legal brief was received from the Governor of Connecticut, protesting that the extent of the contemplated diversions would affect the navigation and self-cleaning capacities of the Connecticut River. He asked that only flood flows be taken from the Swift, feeling that this would be adequate for both his state and Metropolitan Boston's needs. Little attention was given to this brief, although the subject would reach the Supreme Court a year later. [176]

At the hearing on the 8th, a few new points came to light. Davis Kenniston spoke in favor of the bill, and stated that the M.D.W.S.C. had no intention of taking any Millers River water. He was the only proponent of the bill to speak.

Attorney Storrs and John Schoonmaker spoke in favor of several amendments to the bill. Besides favoring more compensation for valley residents, they wanted the towns' taxes to their counties paid for by the M.D.W.S.C. They also sought permanent compensation for Hampshire County for the loss of three towns, being seconded in this by the Hampshire County Commissioners.

Dana representatives pushed for more compensation for the loss of industries in North Dana. They also supported requiring relocation of the railroad east of the reservoir flow area. Annexation of what would remain of Dana to Petersham was opposed by the Dana delegation. They believed the state should support the town until it could rebuild itself. New Salem representatives felt that little would left of their town if the rumored 1000 feet of land next to all watershed streams was to be taken.

Palmer representatives, including Attorney Hobson, felt that not enough protection was given to the remaining flow of the Swift River. Hobson wanted water set aside for a fire district in Palmer.

Athol representatives again expressed support for relocating the railroad east of the reservoir, and asked that the Millers River be left out of the water project. Prophetically, Attorney Duncan asked that provisions be made for policing the project while it was in progress.

N.B. Kimball of the New York Central R.R., lessees of the B & A Branch line running through the valley, stated his feeling that the bill's provisions regarding the railroad were vague. He noted that the branch was currently a paying proposition, but that relocated along miles of reservoir it might not be. Kimball wanted the railroad to be taken by eminent domain in the act.[177]

Roland Sawyer put forth a petition to increase damage provisions in the bill. The hearing then adjourned into an executive session. At this session, and in conferences held the next week, attempts were made to iron out the differences between various amendments.

The Water Supply Committee finally reported a bill on March 16th. The water project in the Swift River Valley would be funded by a $50,000,000 bond issue. Dana was not among the towns to lose its corporate identity in the bill, due to uncertainty about the final height of the reservoir flow. The railroad was to be dealt with by agreement between the M.D.W.S.C. and the line's owners. The employment loss compensation provision was left in. Towns below the reservoir would have future rights to water for fire districts. Hampshire County was to be given $55,000 as compensation for the loss of three of its towns.

No one was entirely satisfied with the bill, but it was seen as a fair compromise by most of those involved in its drafting. The sorest point for the valley residents, the date of the land takings, was not dealt with in the bill, but provisions were made for anyone to sell land when they desired to. The problem of the valley cemeteries was resolved by establishing a new large cemetery near the reservoir for reinternment of valley graves. The bill was referred to the Senate Ways and Means Committee for action. [178]

While the Ways and Means Committee set up a hearing on the bill, new groups were showing interest in the plight of the Swift River valley. Chief Engineer Winsor indicated that no work would commence in the valley until at least 1929; valley people wondered whether or not they should move right away. Members of the South Deerfield Board of Trade toured the valley, trying to get residents to remove to their town, about 20 miles to the northwest. The East Springfield Home Builder's company sent letters to many valley homes with the same thought in mind. Apparently the thrifty valley folk were seen as desirable additions to any town. [179]

The happy lull the water bill reposed in lasted only a week. At the Ways and Means Committee hearing on March 22nd, the bill was sharply attacked by Elijah Adlow, Assistant Corporation Counsel for Boston. Adlow, a former House member, charged that the Swift Valley residents were trying to "drive a hard bargain" with the Metropolitan District. He thought that the unemployment compensation provisions for the valley had a poor precedent on the Wachusett Act of 1896. Adlow made his motives obvious when he pointed out that Boston's water rates would increase to cover the cost of the project, and that he therefore wanted to eliminate "unnecessary" costs from the project. Davis Kenniston and Senator Pond of the Water Supply Committee both replied that the bill was a fair compromise. [180]

Adlow's outburst caused a flood of indignation from valley residents and Metropolitan interests. All parties concerned decide to amend the bill to their own liking, thus killing the spirit of compromise. At the House Ways and Means Committee hearing on April 5th, a raft of amendments came in from all sides. Davis Kenniston brought in 16 from the M. D. W. S. C., while Adlow added 7 of his own. A group of "associated city counsels" from the Metropolitan District supported Adlow. Roland Sawyer asked that all amendments be thrown out, or he would introduce another 18 sent in from valley people. Some valley people spoke against including any new amendments in the bill. 181

As it was in the last month of the session, the General Court had little time to haggle about the bill. This sentiment caused few amendments to be added to the bill, those accepted being mostly of a technical nature. Both branches approved the bill by the end of April 2nd. Governor Fuller signed it, after contemplating a request from the General Court that it change the flow levels for the Ware River as a health measure.

The Swift River Valley was now committed to become the scene of an exodus, although it would take a dozen years to complete.

PART IV   CONSTRUCTION

Chapter One   The Work Begins

The passage of the Swift River Act (Chapter 321, Acts of 1927), with its emergency preamble, gave the M.D.W.S.C. all it needed to get going on the great water project. The commission fleshed itself out as a bureaucratic organization by year's end, employing well over a hundred people. Besides the main Boston office, there were field offices in Holden, Hardwick, Enfield and Framingham. The last was necessary because of the emergency diversion of the South Sudbury River that had been authorized by the General Court.

There was an esprit de corps among the engineers, possibly because of initial resentment toward them by the natives, and their backgrounds. Joke newsletters were produced at the Hardwick and Enfield offices in the fall of 1927, lampooning both the local residents and the work habits of fellow engineers. One of the bits from an issue ran as follows; "That cenetery job will be a pretty stiff proposition in the winter. A little brown jug and a fur lined chair might help." 182

In the first 11 months of 1927, 667 requests for an appraisal of real estate were received by the field offices. These included about 3,500 acres of land that would be needed by the commission. These voluntary offerings were filled out by the property owner on a form supplied by the field offices. These were relayed by the secretary to the chief engineer, thence to the proper division engineer. The latter got an outline sketch and description of the property, with three appraisals attached. Then the commission would act on the chief engineer's recommendation on purchasing the land. If the property owner agreed to the figure supplied, the commission would buy the land. Often, the owner was given the option to rent it back from the commission at a nominal fee. A number of experienced local appraisers were employed by the commission, including W.H. Walker of Greenwich, George W. Cook of Barre, and Walter Clark, who later wrote the book "Quabbin Reservoir".

A complete mapping job was done for all surveys done in the Swift and Ware valleys. This took up much of the first two years work of the division engineers. Because the Ware diversion project was begun in 1927, land purchases there were carried on in a more agressive fashion than in the Swift Valley. As contracts for most of the aqueduct shafts had already been awarded by the end of May, the urgency of landtaking can be seen. 183

The tunnel for conveying water from the Swift Valley and the Ware River was to be over two dozen miles in length. Beginning at a point northeast of Quabbin Lake in Greenwich, it was to run east to a point under the Ware River in southeast Barre. From there, it would turn slightly, and run east to Oakdale, a village of West Boylston at the western edge of Wachusett Reservoir. It was thought that a straight line tunnel due west from Oakdale would have saved some distance, or that a line running from a point further south along the Swift Valley would yield more water, but these were rejected. The final layout allowed for a reasonably short distance, and had the advantage of allowing the Ware water to be run into the Swift Valley at a point far above the intake.

The tunnel itself was to be bored out of the bedrock, then lined with a reinforced layer of cement. About a dozen feet in diameter, it was horseshoe shaped. A special machine was used to spread the cement, one that ran on compressed air instead of electricity. Such safety measures kept the casualty toll down, but six lives would be lost in building the tunnel from the Ware to the Wachusett.[185]

To construct the Ware River intake works and protect the watershed, several villages had to be demolished. Parts of the towns of Barre, Hubbardston, Oakham and Rutland were affected. While these towns did not lose their corporate identity, each lost a village or section to the M.D.W.S.C.

Several brooks in Hubbardston are tributaries to the Ware River, among them the Burnshirt River, Natty Brook, and Canesto Brook. In this town, several mills and mill sites were purchased by the state. Among them were the Wachusett Nail Corp., Hubbardston Chair Corp., the Coffin Mill, the Blanket Mill, and the Jefferson Mill. The Kenan Sawmill was also taken. The village of Williamsville, while on one of the Ware tributaries, was considered far enough upstream and relatively inactive enough to be left intact.

Rutland contains many tributary waters of the East Branch of the Ware River. The villages of North Rutland and West Rutland, being located on these streams, suffered from much dismantling of mills and residences. A state prison camp of almost a thousand acres was purchased for watershed land.

In North Rutland, the Moulton textile mill and adjoining properties were purchased. West Rutland was demolished, with the Rutland Worsted mill and its 54 acres of land going for watershed protection. The schoolhouse there was purchased from the town. Most of the two mile long Long Pond was also obtained by the M.D.W.S.C.

Oakham held several tributaries of the Ware, and that river flowed through the northwest edge of town, near the village of Coldbrook Springs. This was the largest village to be totally demolished by the watershed takings for the Ware River. It had a post office, general store, filling station, hotel and church. The Baptist Church was moved to Greenfield, being one of the few buildings in the Ware River watershed lands to survive. All of the northern section of Oakham was purchased by the M.D.W.S.C. The Ware River Branch of the B & A Railroad had passed by the northwest edge of Coldbrook, with a station serving the village. Forty years later the rail line was taken up, leaving only Mass. Route 122 as a major route through that part of Oakham.

There were (and are) three major villages in Barre; the center, Barre Plains, and South Barre. The latter two, while on the Ware River, were well below the diversion point, and the first is high up on a plateau at the central part of town. While these villages were spared, all of the eastern and southeastern parts of town were taken by the state. That part of the Ware River running through Barre Falls was in this section. At the southeast end of town, a large cotton mill was the center of a village once known as Smithville, then White Valley. When all of this land was taken by the commission, one of the mill buildings was converted into a storage garage, as the diversion shaft was only a few hundred feet to the east.

Over 21,000 acres of land was aquired for the Ware River watershed. While most of this was done by 1931, many outlying tracts were picked up gradually over the next dozen years. Much of this land was eventually opened up for passive and some active recreation.

By early 1928, enough of the shafts and preliminary work had been completed to enable work to begin on the tunnel itself. Two contractors combining as the West Construction Co. won the contract for the work. They employed up to a thousand men in three shifts to do the labor. The M.D.W.S.C. had previously received approval for the diversion from the U.S. War Department, provided that the limits laid down in the enabling legislation were adhered to. [186]

While the Ware River work was commencing, the Swift River Valley saw little action, other than surveys and test borings. Many people did sell out to the M.D.W.S.C. and leave, as evidenced by the drop in the population of the four towns by 30% from 1925 to 1930. Out of the approximately 600 people who left the valley those first three years, almost 200 of them were from Prescott. Although this town had no industry to speak of (which would have triggered an exodus upon closing), and only a small percentage of its land was to be inundated, most sold out within the first year after the Swift River Act was passed. One reason for this could have been the general poverty of the town. As more families sold out and left, there would be more of a burden on those remaining to maintain town services, and more families would become isolated in the remote parts of town. This was especially a concern regarding firefighting, for much of the town was wooded and susceptible to forest fires. The remaining residents also had to be concerned with many breaks into the recently evacuated homes. [187]

At the beginning of 1928, Prescott's remaining citizens asked the authorities for help. Rep. Sawyer, State Senator George Pond, and Davis Kenniston of the commission agreed on a solution. Sawyer filed a bill in the General Court to allow the town to vote to turn over its operation to the M.D.W.S.C. This was not to be the finish of the town as a corporate entity, but it was to be a de facto end. At a hearing before the Water Supply Committee, residents of the town pointed out that they had to support three schools with only 25 pupils between them. They also noted that there were only four horses to plow over 75 miles of roads, and that more people were leaving town soon.[188]

The bill was reported out of committee in early April, and passed at the end of May. At a town meeting on June 25, the few remaining voters agreed to turn over the operation of the town to the M.D.W.S.C. After 106 years as a normally functioning municipality, Prescott was to be run by a half-dozen people as agents of the commission.

While Prescott was the only one of the four valley towns to be rapidly depopulated, the others were experiencing the loss of various businesses and institutions that made it less encouraging to stay on. The Congregational Church in Prescott held its last services in 1928. In Greenwich, the Hillside School closed down to move to Marlboro in 1927. In Dana, Camp Dana and the Mount L Hotel were sold within a year of the passage of the Swift River Act.

Greenwich felt the urgency of the evacuation so much that it held its 175th anniversary celebration a year early in 1928. Reportedly, 5,000 people attended the event, with U.S. Senator David Walsh as the main speaker. One of the main topics of conversation among those attending was the question of whether the state had overdone it s plans for the project. The lack of final plans from the commission, plus conflicting statements from engineers gave people doubts that the whole valley would be needed for years. Connecticut's opposition to the diversions was also cited as something the state should have dealt with before committing itself to purchasing land. [189]

1929 saw more people leaving and more signs of the impending doom that was the valley's fate. Another of Dana's boy's camps shut down. The Swift River Hotel in Enfield closed (later to reopen). Ryther and Warren closed their Enfield store, moving it to Belchertown. A Boston landscape architect proposed that the water project be made into a huge memorial for the First World War dead. Two bills to ease and speed the land taking proceedure for valley landowners were defeated in the General Court.

The next year saw more of the same. The Prescott Hill church was torn down and moved to South Hadley. In April, the Greenwich Plains Post Office closed, following the course of the two Prescott offices which had closed years earlier. The New Salem depot of the B & A Railroad was closed due to lack of business. Samuel Sanford, a 73 year old hermit, was forcibly removed from his shack in Prescott and sent to a state institution. He was in a remote part of town, and suffered from asthma. No one wanted to wade through miles of winter snow to deliver supplies or look after him. [190]

A national depression was just beginning to make itself felt in the region. Massachusetts was celebrating its 300th anniversary as a state. Joseph Ely of Westfield began a campaign that would see him become the first Democratic Governor of the state in a decade and a half. Many of the minds of the Swift River valley folk were focused on something more important to them than any of that. In the Supreme Court in Washington D.C., a decision would be made that represented the last chance the valley had for survival.

Chapter Two   To the Supreme Court

The State of Connecticut had never been wholly in favor of the diversion of both the Swift and Ware Rivers. It had been willing to accept a compromise solution involving only the flood flows of both rivers, but this had been ignored by the Massachusetts General Court at the time of the passage of the Ware and Swift River Acts.

The Connecticut River, as the ultimate destination of the waters of the Swift and Ware Rivers, was an important stream to its namesake state. The river cuts the state in half on its course to the Long Island Sound. It is navigable as far north as Hartford, while numerous farms prosper within its flood plain. At Winsor Locks, above Hartford, various canal and power projects had been built, and others were in the planning stages. While only about 2% of the total flow of the river entering Connecticut was affected by the M.D.W.S.C. projects, the interbasin diversion of the water made Connecticut fight for its riparian rights in the matter.

Riparian law is very basically the doctrine that those owning land along a stream have the right to use it as long as they do nothing to impair the rights of downstream users. This provides for mills to dam streams to use the flow for power generation, as little water is lost in the process. A domestic user, or even a town can remove water for its own use, as most of this returns to the stream via sewage (hopefully treated). In the western part of the United States, where irrigation uses rival domestic uses, the doctrine of prior appropriation of water evolved. This gives the first important water user the right to claim the water in a given area. Connecticut was pleading the riparian doctrine, while Massachusetts was using the unusual (for the northeast) prior appropriation doctrine.

The M.D.W.S.C. had not gone into the projects without some outside supervision. The War Department had requested information on the projects as early as January, 1927. Several months later, it told the M.D.W.S.C. to apply for a permit under its regulations. In May, 1928, after a public hearing, the Department approved the Ware River diversion as proposed, with the provision that no water be taken from the stream between the middle of June and the middle of October. An application was made for the Swift project and approved on May 17, 1929. A formula was set up to determine release of water from the reservoir when the Connecticut River was below a certain level during the summer months. [191]

Even though these safeguards were put into the projects by Federal authority, Connecticut pursued its course of opposition. On July 11, 1927, it asked the Commonwealth of Massachusetts for all pertinent information regarding the water projects. On January 3, 1928, Connecticut filed a motion in the U. S. Supreme Court for leave to file a bill of complaint. This motion was granted, and the bill was filed on January 16.

Connecticut's bill of complaint, as summarized by the court in it s later opinion, follows:

> The complaint alleges Connecticut and Massachusetts recognize the common law doctrine that riparian owners have the right to the undiminished flow of the stream free from contamination or burden upon it. Connecticut appears as owner of riparian lands and of the bed of the river and as parens patriae. The proposed diversion will ... impair navigation, will take flood waters and thereby cause damage to agricultural lands. The Connecticut is now heavily burdened with offensive matter put into the river in Massachusetts and requires all the water that naturally comes down the river to prevent it from becoming a nuisance ... the diversion proposed is only a portion of that covered by the plan ... which includes other tributaries of the Connecticut.

The last sentence reflects Connecticuts later justified fear that diversion attempts would not stop at the Ware and Swift.

Massachusetts filed an answer on March 5, 1928. It amended it s answer when the subsequent decisions of the War Department were in it s favor. In the court's summation:

> By it's answer and amendments thereto Massachusetts denies that the proposed diversion will cause any injury or damage, and avers that the amount of water to be taken is negligible ... that an emergency exists in Massachusetts constituting a justification for a reasonable use of such waters ... that Connecticut's contention that the diversion will interfere with navigation is not open in the absence of proof that Massachusetts has diverted or actually proposes to divert more water than is permitted by the War Department ... that the project will stabilize the flow of the river and result in benefit to Connecticut and lower riparian owners.

Over a year later, in April 1929, Connecticut filed a motion to cite the Army Chief of Engineers and Secretary of War as respondents in the case, and to obtain an injunction against them from acting on the request to approve the Swift diversion. Connecticut alleged that the officials were not qualified to make such a decision. This motion was denied by the court.

On December 2, 1929, Connecticut, with leave of the court. filed a motion to dismiss the defendant's answer and strike out sections of the answer. Massachusetts defended its contentions successfully, as the court denied this motion on January 20, 1930. Connecticut filed another reply to Massachusetts' amended answer a week later, contending that Massachusetts' allegations were wrong, and that there was plenty of water available in Eastern Massachusetts.

The latest formality occurred after the court had appointed a Special Master to take all evidence and report his findings to the court. Charles W. Bunn, Esq. of St. Paul, Minnesota, was appointed by the court to carry out this duty. Bunn held hearings in Boston and Hartford from February 18 to March 21, 1930. The case was also argued before him in St. Paul on April 23-4. The evidence presented at the hearings filled seven volumes totaling 2,540 pages plus a 360 page summary.

Ernest L. Averill, Deputy Attorney General, and Attorney Benedict H. Holden assisted Benjamin W. Alling, the Attorney General, in handling Connecticut's case. Joseph Warner, the Attorney General of Massachusetts, appointed Bentley W. Warren to handle the case, with G.J. Callahan and R.A. Cutler assisting. Many notable witnesses were called to testify for both sides, among them most of the engineers of the M.D.W.S.C. Some bits of the massive amounts of testimony are interesting for the light they shed on some of the men and problems involved in the Metropolitan water supply story.

The testimony revealed that because Allen Hazen had promised a compensating reservoir for downstream users of Ware River water in 1925, many Springfield area millowners had asked for water instead of damages when the project got under construction.

X.H. Goodnough's testimony was quite revealing. When he was reciting the steps leading up to the Ware and Swift projects, Atty. Holden accused him of "rambling around" about the historical aspects of the water supply. To Holden it seemed "a very fine time-killing device".

Goodnough's rather callous feelings about the valley were expressed thusly: "The country that will be covered with water is a very poor soil . . . It is almost sand dunes. Most of the population is in the bottom of the valley which will be covered with water, and that population naturally would not stay within the watershed. It would naturally move out".

As to one of the beauties of the proposed Swift Reservoir, Goodnough saw it as a possible storage place for water diverted from New Hampshire. He always downplayed any possibility of using Eastern Massachusetts sources, stating that Boston would "object very strenuously" to using the Merrimac River. A summary of his justification for the projects could have come when he said "The policy of Massachusetts . . . was to look west for its water supply".

Karl Kennison seconded Goodnough's testimony. At one point, he stated that no Eastern Massachusetts water sources could be developed without "seriously impairing the industrial and other growths" of that part of the state.

J. Waldo Smith, an M.D.W.S.C. consultant, came off very poorly in his testimony. After coming out with a gem like "I've never known one (a reservoir) to be built too large". he noted that it was "necessary" to go to the Swift River. He suffered from lapses of memory, for example when he confused the Ware River with the Delaware. He expressed the opinion that eventually every major tributary of the Connecticut would be stored in the Swift Reservoir, and that Hartford could get water from it. 192

The evidence itself was divided into seven categories: Stream Flow, Navigation, Power, Agriculture, Pollution, Fish Life, and available supplies and demand in Eastern Massachusetts. On the first two points the two states agreed on the facts. As regarded power generation, the special master concluded that the projects would do more harm than good to Connecticut, and came up with a damage figure. He felt that the effects of the projects on Agriculture, Pollution, and Fish Life were very minimal.

On the last issue, the Special Master concluded that because of the long time taken to decide upon the Ware and Swift Rivers as water sources, their practicality and acessability, and poor condition of the eastern state waters, that these projects were justifiable.

Bunn filed his report on May 19, 1930, sustaining the contentions of Massachusetts and recommending the dismissal of the bill. He also recommended provision for protection of the developer of the King's Island project on the Connecticut River be made.

On October 6, 1930, Connecticut filed 75 exceptions to the Special Master's report. Obviously, Massachusetts filed none. Both states submitted briefs and reply briefs in December, 1930 upon Connecticut's exceptions. On January 5 and 6, 1931, both parties argued the exceptions before the Supreme Court for the last time.

Justice Butler delivered the opinion of the court on February 24, 1931. Connecticut's bill of complaint was dismissed "without prejudice to her right to maintain a suit against Massachusetts whenever it shall appear that substantial interests of Connecticut are being injured through a material increase of of the waters diverted by or under the authority of Massachusetts . . ." Each state was to pay its own court costs, plus half of those of the Special Master. No damages were to be granted to any developer at King's Island, Ct., a judgement that differed from the Special Master's. Justice Butler, with an earlier opinion of Oliver Wendell Holmes in mind, noted in the decision that

> Drinking and other domestic purposes are the highest uses of water. An ample supply of wholesome water is essential. Massachusetts, after elaborate research, decide to take the waters of the Ware and Swift rather than to rely on the sources in the eastern part of the Commonwealth where all are or are likely to become polluted. We need not advert to other considerations, disclosed by the evidence and findings, to show that the proposed use of the waters of the Ware and Swift should not be enjoined. 193

There is an important parallel case which came before the Supreme Court at about the same time as Connecticut vs. Massachusetts which is worth noting for the light it sheds on the court's thinking at the time the latter case was decided.

New Jersey vs. New York came about because of the oppostion of New Jersey municipalities along the lower Delaware River to the taking of large amounts of that water by New York City and northeastern New Jersey. A compact was set up between New York, New Jersey, and Pennsylvania to divide the usable flow of the river into water supplies for parts of the three states. This would have reduced the flow of the stream by about a third at a place such as Trenton, New Jersey.

After going through a ser of hearings under a special master, the Supreme Court decided the case in favor of the compact in May, 1931. It did reduce the amount of water the compact could remove from the river, but the domestic uses made of the water by the compact members was seen as the river's most important use. The Delaware River Compact became the forerunner of the river basin commissions of recent years.

Justice Holmes, in his opinion on the case said, "A river is more that an amenity, it is a treasure. It offers a necessity of life that must be rationed among those who have power over it". [194] The Supreme Court in 1963 implied that Congress has the power of apportionment of water supplies between states. As we shall see, the progression in legal thinking about water appropriation will re-enter our story at a future time, but involving the same river basin and opponents.

Chapter Three   Exodus

When the verdict from the Supreme Court came in, the M.D.W.S.C. was elated. The Ware River diversion works were quite complete, and only the legal roadblock held up the use of them. Shortly after the court ruling, on March 20, 1931, the works were turned on. The millions of gallons a day going through the tunnel raised the level of the Wachusett Reservoir by 30 feet in a few weeks.

Contracts were let out and work began on the tunnel from Greenwich to Coldbrook that spring. The inlet of the tunnel was to be on a hillside east of Quabbin Lake, from which it would run under Hardwick and Barre to join the Ware-Wachusett tunnel at Shaft 8.

Since the Ware River landtakings were nearly complete, the towns affected by them were starting to feel the pinch at tax time. Because each building or industry moved for watershed purposes lowered the value of the land involved, the towns were getting less in taxes from the state than previously. A meeting of representatives from many of the towns involved was held in Barre in

February, 1931. John P. Day, selectman from Oakham, noted that his town suffered a $30,–40,000 loss in revenue because of watershed holdings. Representative Clyde Swan of Barre proposed a bill be filed in the General Court to cover all of the problems in this regard. Nothing was done about this until the next August, when such a bill was written for the 1932 session to consider. [195]

While no work of consequence was being done in the Swift River valley, 1931 saw further erosions of the old way of life there. In June, the B & A Railroad announced that it would run only one passenger and one mixed train per day over the valley line, with no Sunday service. This action was protested by some, who felt it would slow up mail service that depended on the train. The railroad also acted indifferently to a proposal to serve Athol via a new spur from Barre to replace the doomed line. [196]

So many homes and cottages had been sold to the state by this time that a method of making use of them until they had to be torn down was developed. Anyone who wanted to rent one of these habitations could do so for $5–30 a month rent to the M.D.W.S.C., depending on the size of the place. A few of the increasing number of tourists to the valley took advantage of the cheap rents, as well as engineers and workers on the projects. Even some of the valley residents who had jobs stayed on in their old homes by renting them after selling out.

The Swift River Hotel in Enfield was reopened when the last operator, Bill Galvin, contracted with the state to operate it. Galvin had been a caretaker at the Dugmar Golf Course in Greenwich after selling out. The hotel prospered, housing tourists, engineers, and official visitors to the project. [197]

Besides the nearly invisible tunnel work east of Greenwich, there were minor projects in the valley to remind the residents of their fate. Late in the summer, work began on a diversion tunnel for the Swift River on the Enfield-Belchertown line. This tunnel would allow the river to flow around the site of the dam while it was being built. Work also began on removing graves from cemeteries in the valley. A new cemetery was under construction in Ware, outside of the future reservoir area, to accomodate the transferred remains. While working at a gravel bank, one crew uncovered a skull from an unknown shallow grave. [198]

One bit of excitement closed out the year 1931. George and Ellis Barlow were brothers who owned a farm near the Enfield-Greenwich line. A long time hobby of theirs was prospecting, which resulted in their discovering gold on their property. Unfortunately, the cost of removing the ore and processing it was far above the value of the metal. The brothers doggedly continued to work their mine, hoping to find the rich vein that would make them wealthly. [199]

In January, 1932, Rep. Clyde Swan of Barre filed a bill in the House to require the M.D.W.S.C. to pay taxes on lands taken for water supply projects at the rate assessed at the time of the taking. After much of the usual acrimonious debate on the subject, the bill was passed two years later.

Work on the tunnel connecting the intake shaft at Greenwich and the Ware River continued through 1932, with 300 men employed on it at one point. The accidents and deaths that marred the construction of the eastern tunnel were reduced by added safety measures, with special employees to check on them. [200]

Other work going on at the time included more cemetery removals, and a prohibition of further burials in valley lots in August. Caissons were being pored as the first stage in the construction of the dike to be built over Beaver Brook at the Enfield-Ware line. Surveys were made of routes to bypass the future reservoir on the west and east sides.

Among the property questions that came up during the year was the fate of the valley rail line and the Dugmar Golf Course in Greenwich. The railroad was the subject of meetings between its owners and the M.D.W.S.C., but nothing was resolved that year. The golf course, bordering on Curtis Lake, had started out as a retreat for Thomas Maher and John Duggan of the Chapman Valve Co. of Springfield. It developed into a private nine hole golf course and country club by 1927. Although rather sandy, the course saw some use by the local populace. No agreement was reached by the two parties in this case, which would lead to extensive legal proceedings. [201]

Remembrance of the past and plans for the future were also on many minds at this time. At an old home day in Prescott in September, 1932, it was suggested that a historical society be formed to perpetuate the memory of the town. This was done, and New Salem formed a society at almost the same time. Members of both groups met later in the year to work on forming a society for the whole valley. Because the Prescott people wanted each town name to form a part of a cumbersome name for the society, its formation was delayed for another four years. [202]

Various groups were proposing future uses for the vast reservoir and its watershed. The Forbush Bird Club of Worcester wanted to turn future islands in the reservoir into bird sanctuaries. A few months later, the Metropolitan Improvement Association, a high flown Boston group, wanted to name the reservoir after Calvin Coolidge, who had died in January, 1933. The watershed lands were to be named Calvin Coolidge Memorial Park, and the road from the Brookfields to Northampton (running south of the reservoir) Calvin Coolidge Highway. None of these proposals got past the agitation stage. As for the name of the reservoir, the M.D.W.S.C. voted to call it Quabbin Reservoir in October, 1932. This was probably due to the historical researches of engineer Karl Kennison, who noted the old uses of that name in the locality. [203]

The nation was deep in the throes of the great depression in the early months of 1933. Massachusetts was as hard hit as everywhere else, but the valley did not seem to be as affected by it. Most of the people there still had jobs or were surviving on their farms. Even though land was hard to sell in many places, the M.D.W.S.C. was still buying what they needed at reasonable prices for the time. Because of this, some offers to sell were received from landowners outside of the contemplated reservoir area. The state ended up paying an average of $103.64 per acre for real estate (including buildings, etc.) in the Swift Valley, while the average cost in the Ware Valley had only been $83.45 per acre. [204]

For the valley, one of the most important single events of early 1933 was the formal dedication of the Quabbin Park Cemetery. Work had been done on the cemetery since land for it had been acquired in 1931. Of the 104 acres of land, only 13 were used for burial plots. 34 Cemeteries in eight towns had to be relocated to Quabbin Park, Eventually, 6,551 bodies were relocated there, while another 1,101 were buried in other cemeteries. [205]

Quabbin Park Cemetery was beautifully landscaped, and located just off Route 9 south of the dike site. A handsome building housed all of the records of the original and current lots. Narrow gravel street ran between rows of well placed markers. Many markers and memorials op public significance from the valley towns were placed in the cemetery. Unfortunately, the graves were not placed with any consideration of their location in the original cemetery, or even by town. This gross oversight marred an otherwise tastefully carried out project.

When carrying out the task of moving graves, the M.D.W.S.C. attempted to notify next of kin or decendents of each of the deceased, to allow them to be present at the reinterment in the new cemetery, or to arrange for relocation in a cemetery of their choice. Most of the one seventh of the total number of graves relocated outside of Quabbin Park were done in the early years of the work. Many of these ended up in nearby towns, such as Athol, Orange, Belchertown, and Ware. [206]

While the dead were being removed the living were moving out. In Dana, the tax rate showed a large decrease from the previous year, and the town was free of debt. However, it was noted that fewer than two dozen homes remained under private ownership in the town, with the commission having torn down a few others already. [207]

Depopulation was becoming more noticable elsewhere. The congregation of the 80 year old Enfield Methodist-Episcopal Church voted to close down the church the next spring. Proceeds from the sale of the edifice to the Commission were donated to the Fairview Methodist Church. Of the taxes collected in Enfield that year, over half came from the M.D.W.S.C. Prescott had no problem being the first to pay it's county tax for the year; $190.72. The tax rate in the town of New Salem reached $33.69, almost twice the previous year, but this was blamed on generous town meeting appropriations. [208]

Work was progressing on the bypass highways around the reservoir area. The highway on the west side, connecting Belchertown and Orange, was causing some grave shifting. It seemed that at Pelham Center, the cemetery had to be partially moved to accomodate the highway, or it would have a dangerous curve and force out homes. Graves were transferred to the other side of the cemetery from the highway bed, and a knoll was built to separate the highway from the burial ground.

The owners of the Dugmar Golf Course sued the M.D.W.S.C. for $436,500 in damages in October. This case was brought before the Board of Referees established for that purpose. Previously, a six figure settlement had been received by the owners of Camps Quabbin and Pomeroy in Greenwich. When the commission offered only about $50,000 for the golf course, the owners resorted to court action. The board awarded the owners $221,000 for the property, but the M.D.W.S.C. appealed that figure. [209]

Work on the Coldbrook-Greenwich tunnel continued to within 100 feet of completion by the end of 1933. A strike slowed down work for a while in the beginning of August. Eventually, 280 men walked off the job in a bid for unionization and higher wages. After an appeal to the Governor, an arbitrated settlement was reached, with guarantees of better conditions for the men and a minimum wage of 40 cents an hour. This was the first of several different labor problems that would slow the reservoir project only slightly in the next half decade. [210]

The year 1934 saw more work on the project in the valley commence, although it was still on a small scale. The core wall on the dike was completed, so that in late November, the A.A. Johnson Corp. was awarded the contract to build the dike embankment. Excavations were done to find fill for the dam near its site. The New England Power Co. moved one of its transmission lines further west so that it would go around the reservoir site. By the end of the year, the M.D.W.S.C. had purchased over 60,000 acres of land in the Swift River, or about three-fourths of what it would ultimately take. [211]

A parallel project was completed that year, the Rutland-Holden sewer line. This had been made necessary by the potential pollution of Metropolitan water sources in those two towns. The sewer, opened early in November, channeled the town's wastes through the Worcester system. M.D.W.S.C. personnel were transferred from the closed Holden office to those in Barre and Enfield. In August, Worcester set up a pumping station on the Quinnepoxet River to augment its water supply. This was based on the 1926 act allowing that city to have a share of the watershed of the Wachusett Reservoir.

Court problems continued to plague the M.D.W.S.C. Five settlements were made that year with injured parties on rivers below the projects. The largest was to United Electric Light and Indian Orchard Co. of Springfield, who got $375,000. As already noted, the owners of the Dugmar Golf Course were awarded $221,000 by the Board of Referees. The commission appealed this ruling to the Supreme Judicial Court, while people played for free on the abandoned course. Politicians in Boston thought that the case had been handled unfairly towards the Commission, as all three of the members of the Board of Referees were from the Springfield area. A bill was passed requiring that one member of the board be from the Metropolitan District. A Malden lawyer replaced a Springfield lawyer on the board as a result. 212

The major settlement question of the year was the Boston and Albany branch railroad. In April, the last regular passenger train had been dropped from service, leaving only one mixed train going each way each day. During the summer, the New York Central Railroad, the owner of the B & A, filed an application with the Interstate Commerce Commission and the Mass. Department of Public Utilities to abandon the line north of Bondsville, in Palmer. The state hearing was held in September, with representatives of the towns of Athol and New Salem protesting the lack of a relocation. Davis Kenniston and Frank E. Winsor supported the counsel for the railroad in pushing for a total abandoment, probably because the M.D.W.S.C. did not want to pay for the higher cost of a partially relocated line. The opposition by the towns was to no avail, as both state and federal approval for the abandonment came by year's end. Although neither of the parties involved wanted it made public, the figure agreed upon for the settlement to the railroad was released during the public hearings. It was $575,000, with the railroad being allowed to remove the tracks and rails after discontinuing service. 213

Many curious things happened in the valley during the year. In May, a group of nudists tried to buy or rent some land in the eastern part of Greenwich. Although the nudist "craze" was manifesting itself in different parts of the country at this time, the M.D.W.S.C. discouraged them from setting up a camp. In the September political primary, none of the 15 voters in Prescott cast a ballot. A month later, two men were arrested for stealing dynamite from a contractor's depot in Gilbertville in 1932. This was one of a string of thefts that would occur as more equipment was brought in for the project. A fire of unknown origin destroyed three buildings at Shaft 10 in Hardwick in December. A $4,000,000 proposal to build a highway and bridges across the reservoir was dropped as being unecessary. 214

An important change in the makeup of the M.D.W.S.C. occurred at the end of 1934. Governor Joseph Ely was being succeeded by the notorious James Michael Curley, a fellow Democrat in name only to Ely. Ely made several last minute appointments to various state positions, feeling that his choices would be superior to those of his successor. One of these was Eugene C. Hultman to the post of Chairman of the M.D.W.S.C. Curley protested that Hultman was a "civil engineer and a Republican", and thus unfit for the post by his standards. 215

Curley's description of Hultman was not inaccurate; the latter was a Republican, and an M.I.T. graduate who had done some consulting work in engineering. Hultman had also served as a Boston Fire Commissioner, in both branches of the General Court, and as a reforming Boston Police Commissioner. In the early 1920's, Hultman had been the state Fuel Administrator while serving on the Commission for the Necessities of Life. He had been charged with conflict of interest at the time, but had been cleared. 216

When he assumed the Governor's chair in 1935, Curley had Hultman called before the Governor's Council on charges that Hultman had sampled a bottle of confiscated champagne and spread manure from city stables on the lawn of his summer home. The travesty filled 1,100 pages of testimony before it petered out. Curley's motives for trying to remove Hultman became more obvious later in the former's term: Hultman stood in the way to total control of the big contracts and numerous jobs that the water supply projects represented. 217

The year 1935, saw the opening and naming of the long highway running west of the Swift valley connecting Orange and Belchertown. It was christened Daniel Shays Highway after the insurgent of 150 years earlier. The old Conkey Tavern on the West Branch of the Swift, which Shays frequented was in ruins, as was his nearby home, but his memory was still honored.

More work commenced on the dam sites. In August, roads running through the main dam site were discontinued, so that work could begin on the core wall. This contract was awared to C.J. Maney and B. Perini & Sons, operating as the Two Companies, Inc. Various diversion channels near the intake and above it were also under construction. The dike embankment began to rise above the little Beaver Brook valley.

In March, Dana reported it was debt free for the third year in a row, although the state census of that year reported that it was the least populated town in Worcester County at 387 souls. This indicated that over 200 people had left the town since 1930. Greenwich and Enfield had fared better, probably due to the number of residents in these towns who found work on the project, and the number of homes rented out by the commission. Enfield lost a net of two people in dropping to 495, while Greenwich lost only 19 to go to 219. Prescott, already a ghost town of 48 people in 1930, was down to 18 five years later, these mainly being members of three families still living there.

The Greenwich Congregational Church joined the list of valley institutions passing into history, but it left a legacy. In March, the Ware Congregational Church accepted the Greenwich Foundation, which was set up by the Greenwich Church to commemorate itself and provide funds for youth work. The initial $23,000 funding came from the sale of the church building to the commission. The foundation was dedicated in a formal ceremony that October.

One of the more touching events of the year was the last running of the so-called rabbit train through the valley on June 1st. The event was well publicized among rail buffs, who filled both an earlier "last run" and the real one.

115 passengers made the last ride, including Otis and Arthur Hager, two brothers who had ridden the first train on the line over 60 years earlier. John D. Smith of Athol watched the last train arrive and depart from Athol, as he had seen the first one do. There was much autographing of timetables and postmarking of covers for philatelists.

The train had carried people to work, school, and shopping. It had shipped out the products of mills in North Dana and Enfield, and ice cut from ponds in Greenwich and Millington. A work crew spent July and August tearing up the tracks from Athol down to the dam, and the line was no more. One of the last industries to leave the valley, the Swift River Box Co. of North Dana, moved to Athol in July because of the loss of the rail connection. The state delayed paying for the railroad for several weeks, because Governor Curley wanted an investigation to determine how the settlement figure was so large. 218

Other events that year included a forest fire in Prescott that levelled about 2,000 acres of timber. It was thought to have been set, as five different small fires were seen to converge into a single blaze. Governor Curley visited the valley in June, perhaps dreaming about ways he could (and would) use the project to give jobs to potential voters. The Board of Referees award to the Dugmar Golf Course was set aside by the Supreme Judicial Court. This enabled the M.D.W.S.C. to get a new hearing on the settlement.

In 1935, X.H. Goodnough died at his summer residence in Maine. After retiring from the Health Department in 1930, he had formed a consulting firm named after himself, and continued to act as a consultant for the M.D.W.S.C. Goodnough published a glowing tribute to his mentor Dr. Walcott upon the latter's death at 94 in 1932. In spite of what many hailed at the time to be his great accomplishments, no monument or memorial would be erected to Goodnough for 30 years.

1936 became the first year of highly intensive construction work in the valley. Many projects were going on and started at the same time, and no remaining resident escaped the sight or sounds of the work that year. The makeup of the dams and other construction projects will be detailed in the next chapter.

Chapter Four   The Dams

When the plans were made for the Quabbin Reservoir, it was noted that two dams would be necessary to hold back the waters of the Swift River. The The main dam would be located across the valley of the Swift on the boundaries of Enfield, Ware and Belchertown. Another dam had to be built just north of the Enfield-Ware line east of the first one. This was necessary to block off the valley of Beaver Brook. Because there was no spillway at this second structure, it was always referred to as a dike.

Four other smaller dams were also built as a part of the reservoir project. At the southern end of Mt. Zion, near Greenwich Village, two stone embankments were raised, running north to south. When the reservoir was filled, these connected the island of Mt. Zion to the mainland near the intake shaft. Their purpose was to divert the flow of the East Branch of the Swift and any introduced Ware River water around Mt. Zion. Since the latter two waters were considered inferior to the other sources, this gave them much more time to clean themselves before getting to the intake. These dams were known as baffle dams because they changed the direction of the water flow.

Two small dams were built in the closing days of the project at the upper reaches of the Middle and East Branches of the Swift. Original uncertainty about the final elevation of the reservoir had left some doubt as to its final extent in these areas. When the elevation was set at 530 feet above sea level, this was not quite high enough to flood these areas without leaving them to form swamps. To prevent this, small horseshoe shaped regulating dams ensured that the uppermost reaches of the two branches would have sufficient flowage.

Of the six dams required for the project, the first two were the most important. They had to be specially built to hold back 412 billion gallons of water. More than two years were spent building them, along with millions of dollars, and losses of life. Job actions, investigations, and court suits would also mark the progress of work on the two main dams.

Before any work could be done on either of the main dams, several years were spent investigating the soils and bedrock at the two sites. It was determined that an impervious core of caissons would be built from the bedrock to the surface of the land at each site. On top of this the dams would be formed by hydraulically filled earthen embankments. Sufficient amounts of acceptable soils for this purpose were located near the dam sites.

Beaver Brook was too small to pose much of a problem in diverting it while the dike was under construction. The Swift River, on the other hand, required a diversion tunnel, which was completed in 1933. Work on the 37 dike caissons commenced the same year, while the 40 dam caissons were not begun until 1935.

Politics began to enter the dam projects at an early point. Late in 1932, M.D.W.S.C. Commissioner Thomas Lavelle objected to the award of the dike core contract to the West Construction Co. Although West Co. submitted the low bid of just under $1,000,000, Lavelle felt that they were "aliens" and should not get the job. The fact that many of West Company's employees were Canadian did not bother the other two commissioners. [219]

The core contract for the dam also had its political side. Two contractors, C.J. Maney, Inc., and B. Perini & Sons, Inc., operating as Two Companies, Inc. won the contract for that job for $1,000,000. Subsequent investigations by the U.S. Treasury Department found evidence that $80-100,000 of that sum had been kicked back to a politician as a bribe. Four officers of Two Companies were convicted for income tax evasions and fined. The "politician" was not publicly identified, but it must have been someone who at least had access to the bids submitted for the project, thus tipping off the Two Companies officers on what amount to submit in their bid. 220

Two Companies, Inc. was involved in another bidding controversy in 1936. Shortly after the Benjamin Foster Co. had been awarded the dam embankment project, two of the M.D.W.S.C. commissioners decided to award part of that work to Two Companies. The latter company was to receive three times the amount that Foster Co. would have received for the same work. It appears that Commissioner Lavelle (who had earlier objected to the Foster sward with the approval of Governor Curley) got the other Democratic member of the M.D.W.S.C. (newly appointed Edward J. Kelley) to a-long with a contract award to a "pet" contractor. The Governor's Council would not approve the contract award to Two Companies, going along with the objections of M.D.W.S.C. Chairman Hultman. 221

Probably the most political aspect of construction work at the reservoir came in 1936. That year it was decided to carry out a project of clearing scrub and brush growing in areas below the future flow line of the reservoir. Since many of these low areas did not have enough merchantable timber on them, work crews would have to be hired by the commission to do the job. The chance to control so many jobs was not to be missed by Governor James Michael Curley. As a long time political boss and former Mayor of Boston, Curley found a way to get many of the jobs for his city constituents.

Curley got control over the hiring by his control of the Governor's Council. Councillor Dan Coakley of Boston worked with Curley and John T. O'Day of the Governor's office in hiring their favorites. Many legislators began protesting this setup as early as April, 1936, but to no avail. The power of these men to hire was so well known that a joke came out of it: the initials M.D.W.S.C. were interpreted to mean "Men Desiring Work See Coakley". 222

Eventually, over 3,000 men were hired to work on the clearing project. The often inexperienced city boys did not prove to be very efficient; charges were made that they did only about one quarter of the work a local woodcutter could do in a day. Tools were stolen or "misplaced", some to be found after the job was over. One man died when a tree fell on him. Such efficiency earned the laborers nicknames like "woodpeckers", or "Curley's Thugs". 223

The influx of so many men became a burden on such nearby towns as Palmer and Ware. The crime rate in these towns increased, mainly in the categories of drunken and disorderly conduct. The Ware police chief bemoaned the estimate of half of the workers boarding in his town. Since there was no public transportation for the workers, vehicles being overloaded with commuters became a traffic problem. Boston legislators charged that their constituents were being overcharged for food and rent by the residents of Ware and Palmer. The Palmer Chamber of Commerce denied this after an investigation. [224]

Other aspects of the clearing project caused local resentment. Because of the large number of men working in the basin area, hunting was prohibited on most of the M.D.W.S.C. lands after October. The loss of a good game territory rankled many local hunters. Protests were made to reinstate the privilege, but it was never done. Another irritating event was the granting of a half-day holiday to the workers on state primary day. This was to enable them to go to Boston and vote for Curley in his senatorial campaign. Few took advantage of the holiday for the intended purpose; bus fare to Boston was $4.00. [225]

The clearing project came to an end in January, 1937. Most observers judged it a failure, except for the fact that it provided temporary jobs. Much of the cleared area had to be done over again by a more professional crew two years later.

The construction of the dam and dike was, on the other hand, a technically complicated affair. After the caissons were sunk across the valley floor to form an impervious core wall, concrete side embankments had to be built. These ran up the hills on both sides of dam sites like a series of steps. Finally, the embankment of earth actually forming the dam could be built. This involved a conveyor belt system to bring the fill (mixed with water) to the proper spot on the embankment. The fill was dropped off, and the water allowed to drain off, causing the fill to settle and harden. Over 2½ million cubic yards of fill were used to build the dike; over 4 million to build the main dam. The process of excavating the fill and sluicing it over the dam sites took about two years in each case.

While the dam and dike were under construction, reforestation was being carried out on many sections of the Swift watershed above the projected flow line. Millions of seedlings were planted over old fields on hillsides and plateaus. Most of these trees were red pine, which would later prove to be a less desirable species than others for the purpose. Nurseries were set up in Greenwich and Belchertown, the former being discontinued in 1939.

1937 saw the dike near completion, and the dam start to rise over the Swift River. Other than these events, it proved to be a less exciting year than the previous one. Lawsuits were still being tried over aspects of the reservoir project. The Dugmar Golf Course received a scaled down award of $150,000 plus interest on the appeal by the state. The Collins Co. of Wilbraham got $300,000 plus interest for water damages. Three Hardwick men who had sued the state, contending the Greenwich to Barre aqueduct drained a brook running through their property lost out in the decision. [226]

Law enforcement activities in the reservoir area were much simpler than in the year before. Men from Ware and Three Rivers were caught stealing dynamite from the dam supplies; they were going to sell it to local farmers. Disgruntled hunters continued to protest the prohibitions on the sport in the reservoir area. The M.D.W.S.C. had been a bit lapse in posting no trespassing signs, and waited until the first day of deer hunting season to do so. Several hunters who entered unposted land in the morning were greeted in the afternoon by policemen standing next to No Hunting signs. These incidents could have been blown out of proportion, but the men were let off with warnings.[227]

The large sums of money now being paid out in connection with the projects were drawing the attention of the Governor's Council. They were becoming apprehensive about the large figures they had to approve, and wondered if the whole project was worth it. Five members of the Council and the M.D.W.S.C. visited the reservoir site on November 15th. Many of the Councillors came away impressed with the "expanse and magnitude of the undertaking". This attitude would prevail until a few months later, when disparate groups would call for the stoppage of all work on the project. [228]

By the end of the year, the M.D.W.S.C. was making new attempts to clear out more of the timber from the valley bottoms. Fifteen sawmills were cutting trees on contract by December. This and other work was going so well that in January, 1938, Commissioner Hultman was predicting that flooding would begin within nine months.

March 1938 saw eruptions of protest against the cost of the reservoir from Boston. Mayor (later Governor) Tobin testified before a legislative committee that Boston's water costs had doubled while consumption stayed the same over the last 20 years. He wondered if the project could be slowed down or stopped to save money. This feeling was shared outside the city, as a Lowell paper editorialized the same thoughts a few weeks later. [229]

Later in April, the Boston Municipal Research Bureau asked the legislature to stop spending by the M.D.W.S.C. It also made several suggestions to improve the Metropolitan District arrangement. Thaddeus Merriam, a consulting engineer for the M.D.W.S.C., and Eugene Hultman protested suggestions to halt Quabbin work, noting the progress of the dam. Although Mayor Tobin supported the Bureau's contentions, no legislative action was taken to stop the project. [230]

Whether people favored the project or not, the rise of the dam embankment drew hordes of curious visitors. A reported 30,000 people came just to see the dam by mid-June. 150 people also came daily seeking jobs. Few could be hired, even though construction of administration buildings just west of the dam was underway. [231]

To assist the unemployed (and probably help his chances in the upcoming primary), Governor Hurley proposed another clearing job in the Swift valley be carried out by the M.D.W.S.C. This was approved by the commission in August, but the work did not start until October due to a dispute about what to pay the workers. Although the top wage for such work in the valley area was about 55 cents an hour, Chairman James Moriarty of the state labor department ruled that it would pay 62½ cents per hour. The state also ruled that half of the jobs would be filled by Boston area men. Some funding was received from the Federal Public Works Administration for this and other M.D.W.S.C projects. To insure that a proper job was done, the clearing work was contracted out by the Commission to two companies: Coleman Brothers and C & R Construction. They had the right to hire and fire by qualifications and amount of work done. [232]

Just before the clearing project began, the great hurricane of September 1938 struck. A flood in March, 1936, had damaged some bridges and a few structures in the valley, but that had been mild compared with what this storm did. Storage sheds at the dam were blown apart, as were some of the houses still standing in the partially abandoned valley floor. More bridges were washed out by the ensuing floods, but the dam, dike, and associated structures held fast. Great quantities of timber were blown down, although much had been removed in the past few years. The Foster Company, contractors at the dam, won one of the bigger contracts handed out by the state for cleaning up hurricane damage. [233]

As 1938 came to a close, the dam was being topped off, and a contract was awarded to place topsoil on the embankment's downstream side. A strike call by the business agent of one of the unions involved in the brush clearing project fizzled out. The agent asserted that Springfield men were the last hired and first fired. The contractor answered that the Springfield men had proved to be incompetent. [234]

1939 began with a shocking event - - the death of Chief Engineer Frank E. Winsor on a witness stand in a courtroom. Winsor had been testifying in a hearing on a suit brought against the Commission by the A.A. Johnson Co. who had built the dike embankment. The Johnson Company charged that they had to spend more than predicted on the job because of faulty information on soil tests and borings supplied by the engineers. The suit was brought in Suffolk Superior Court, and Judge F. Delano Putnam named master.

Attorney Francis Leahy, counsel for the contractor, produced a copy of what was purported to be a memorandum by Winsor ordering the substitution of false data for the real information. When the paper was produced while Winsor was on the witness stand, the latter agreed it contained his signature. Leahy asked Winsor to read it. After Winsor turned pale, but did not reply, Leahy repeated the question. At that point, the engineer suffered a fatal heart attack and died on the stand. [235]

Winsor had cut short a vacation to appear at the hearing, and he was 69 years old. While he must have been under some physical strain, he had testified in court cases before. The decision of the judge would appear to confirm some guilt on Winsor's part, as an award of $285,000 was made to the contractor in 1940. However, that decision was overturned on an appeal to the Superior Court in 1941, and was successful when the contractor lost an appeal to the State Supreme Court in 1945.

In honor of the late engineer, the Commission named the main dam after him. A small park overlooking the dam on the side of Quabbin Mountain was dedicated to Winsor's memory in 1941. The engineers bought a plaque with his portrait on it which was mounted on a block of stone.

The Foster Company brought a similar suit against the M.D.W.S.C. This case involved the engineer's supervision of the contract work, and the location of borrow pits for the dam fill. A court auditor found for the Commission, and was upheld by a higher court. Even though the Commission lost many cases involving real estate settlements, it left a clean record as far as its engineering practices were concerned.

In spite of the court problems, the work had to go on. Karl Kennison, who had been an engineer for the M.D.W.S.C. from the beginning, succeeded Winsor as Chief Engineer. Bulldozers were knocking down trees and piling them up to be burned in the valley. Preparations were being made to plug the diversion tunnel for the Swift River, as 1939 reached its half way point.

Roland Sawyer filed a bill in the General Court to study the feasibility of using Quabbin Reservoir to store floodwaters of nearby rivers during emergencies. During hearings on this measure, Eugene Hultman stated his opinion that the Millers River would be the next stream to supply Quabbin, but not for a long time. Later hearings in Athol saw local opposition to the proposal as a backhanded way of adding water to the reservoir. Flood control dams planned for the Millers and its tributary Tully River were seen as adequate for flood control. [236]

The summer of 1939 saw the completion of the brush and debris removal in the lower parts of the valley basin. All of the people living there had been removed. All roads leading into the valley were blocked off with guards posted. The dam and dike were ready to hold back the waters of the Swift River. On August 14th, 1939, the diversion tunnel was sealed, except for a portion that would allow the 20,000,000 gallons a day to flow through the dam. At a rate of 500,000 gallons a day, storage in Quabbin Reservoir had begun. [237]

Chapter Five   Four Towns Cease to Exist

As we have seen, 1936 was the first year of intensive construction work in much of the valley. Although this activity was pervasive, it did not stop the remaining valley folk from going about their everyday business. 104 voters in the town of Dana went to the polls to vote in the September primary; old traditions were upheld as 87 of them cast Republican ballots. [238] Civic Clubs and Masonic lodges were still functioning, often with engineers and their wives bolstering the membership rosters.

Buildings were being torn down with increasing frequency. The few remaining structures in the Ware River valley watershed were taken down. The most notable of these was the Coldbrook Baptist Church, which was reassembled in Greenfield. Another church which felt the wrecker's hammer was the Greenwich Congregational. It's bell was salvaged and sold to the Polish Catholic Church in Bondsville. One of the last reminders of the valley rail line disappeared with the dismantling of the Enfield depot building in September.

The M.D.W.S.C. took two approaches in dealing with building removal. Some of the more attractive or valuable structures were auctioned off to the highest bidder. Many of these buildings were carefully taken down and removed to be rebuilt elsewhere. Several dozen buildings survived in this manner, most notably the Field House of Enfield, being moved to Dorset, Vt.

Most valley structures suffered a less glamorous fate. Contractors were allowed to tear many buildings apart and cart the material away for their own uses. The poorer structures and remains of the others were just bulldozed into the foundations. Sometimes years elapsed after houses were abandoned before they were demolished. This made for many a spooky sight in the valley. [239]

The onset of spring in 1936 brought an unusually high amount of seasonal runoff into the local streams. This caused extensive flooding in many parts of New England, which the valley did not esecape. The Swift River attempted to breach the embankment protecting the dam site, but it failed. Still, many roads and bridges were damaged, some being discontinued due to high repair costs. The town of Dana suffered $29,000 in damages. South Barre avoided some of the flood when the Ware River was diverted to Wachusett Reservoir, however the latter was filled before the flood waters receeded, so the diversion had to be stopped. [240]

Summer brought one of the most dramatic events in the valley's history. On August 1st, nine days short of it's 150th anniversary, the beautiful Enfield Congregational Church burned down. An adjacent home and garage also succumbed to the flames. Although it was never proven conclusively, the fire appeared to be one of a number that had been set in the lower valley early that morning. Total estimated damage was set at $75,000. Everyone from disgruntled valley residents to discharged "woodpeckers" were blamed for the conflagration. Some, like Reverend J.C. Andrus, saw justice in the event: "It is better that the Enfield Church should go up in a blaze of glory, than to have been torn down stick by stick". Since the M.D.W.S.C. had owned the building for a couple of years, such sentiments were not hard to come by. [241]

In spite of the fire, the congregation celebrated the sesquicentennial of the parish in the town hall on August 9th. 200 people attended the ceremony, marked by Randolph Merrill's sermon. He expressed satisfaction that the church edifice was "safe from possible desecration". 242

1937 saw no spectacular events occurring in the valley. It was another year in the series marking time in the valley's decline. In the fall, remaining valley farmers were told not to plant anymore crops as a sanitary measure. Two notable valley hostelries were torn down, the Eagle House in Dana and the Swift River Hotel in Enfield. Property owners and renters were warned in September that they would have to leave the valley by June, 1938. Like a similar warning made ten years earlier, this prediction turned out to be a bit premature. 243

1938 turned out to be the climactic year in the history of the Swift River valley. It was in this year, on March 28, that the Commission formally took over the whole valley by eminent domain. The four towns also lost their legal existence, and almost all of the man made landmarks in the four towns were gone by the year's end. These events caused the newspapers to devote a great deal of attention to the dying valley - - who could resist the human interest aspects of the events?

In January, property owners in the valley were sent another notice asking them to vacate by April 1st. This caused some to move out, but this notice was not final, as those with children in the local schools or other good excuses could stay on longer. There were enough people left in the three larger towns to hold final town meetings for them in the spring. 244

Dana was the first of the towns to hold its final town meeting on March 7. The few voters present elected the usual town officers to one, two, and three year terms, ignoring the eventual demise of the town. A sum of $25,765 was voted to operate the town, which would end its existence free of debt. Mr. Fred Doane received the thanks of the town for his thirty years of service as town clerk.

Enfield was the next to hold it's last meeting, on April 9th. Even though a sleet storm raged that night, 26 of the town's 30 remaining voters showed up. They were outnumbered by the press and the curious. The noted Dr. Segur called the meeting to order; Postmaster Edwin C. Howe was named moderator. The last of the five articles, which was approved by the assembly, appropriated $1,800 for the construction of a monument to the town's World War I veterans. A committee of five was named to arrange for the erection of the memorial in the Quabbin Park Cemetery.

Greenwich was the last to hold its final town meeting on April 21st. The press covered this meeting in full force, and cajoled the Selectmen of the town to hold a pencil eraser on a globe, among other staged symbols of the town's demise. The old town hall, currently being used as a school, was the site of the meeting. The four article warrant was moderated by Lewis Johnston, whose wife, Lillian, recorded the proceedings as town clerk. Mr. Johnston amused the outsiders attending the meeting by his using his dance prompting vernacular in running the affair. Mr. Charles Walker was appointed to select a tree from the town forest to be placed and marked in the Quabbin Park Cemetery in memory of his brother Stephen, who had died in World War I. $2,000 was appropriated to purchase and erect a memorial to the town and it s war veterans in Quabbin Park. Thus ended Greenwich's 184 years as a town. 245

In spite of all these determined preparations, the state did not get around to ending the towns' legal existence until the end of April. A bill was filed in the General Court to legally end the existence of the towns and transfer land to adjoining municipalities. This was passed and signed by Governor Hurley on April 25th, to take effect at 12:01 a.m., April 28th.

This law provided for the annexation of parts of Enfield, Greenwich, and Prescott to New Salem; of parts of Enfield to Pelham and Belchertown; of parts of Enfield and Greenwich to Ware; of parts of Greenwich to Hardwick; and parts of Prescott and Greenwich and all of Dana to Petersham. Hampshire County lost territory, while Franklin and Worcester Counties gained some. The actual gain of land by these entities was merely geographical, as almost all of the land of the four valley towns was owned by the Commission. No taxes or income would come from the parts of the four towns to their new municipalities for over three decades. Hampshire County was awarded a sum of $55,000 as compensation for the loss of three towns.

To unofficially mark the end of the four towns as legal entities, the Enfield firemen put on a great ball in the town hall building on the night of April 27th. Almost 1,000 people attended the event, which put a great deal of strain on the building meant for a quarter of that number. Former Enfield policeman Coolbeth was pressed into service to direct traffic, but the jam proved too much and he had to be spelled by men from the Ware police.

The orchestra and dancing were stopped just before midnight by Dr. Segur, who chaired the event. When the clock struck twelve, much sobbing was in evidence as the towns officially passed out of existence. This event may have been the height of all the sentimentality attached to the end of the towns, as shown by it's great attendance and newspaper soverage. Some wag observed that the people were now in a town hall that was no longer in a town that was no longer. 246

People still remained in the towns; there were schools to wind up, organizations to end, and jobs for the Commission to be done. The Quabbin Club, the women's club in Enfield, closed its affairs with a large gathering in Ludlow on April 12th. The Zion Chapter, Order of the Eastern Star, went out of existence in Enfield on April 28th. The Enfield and Greenwich Granges gave up their charters on June 16th. The Bethel Lodge of Masons, founded in Enfield in 1825, ended it s existence in 1939.

Final Memorial Day excercises were held in Enfield and Greenwich on the appointed day. A larger ceremony was held at Quabbin Park Cemetery, beneath the tree planted as a memorial to Stephen Walker of Greenwich.

The last official act of the towns was the closing of their grammar schools. This occurred in June, 1938, with each school holding appropriate ceremonies. Enfield graduated seven students on June 22, being the last of the towns to do so. At each ceremony, speeches were delivered, poems were read and hymns sung. The towns had always sent their high school age children to high schools in Athol, Orange, or Belchertown, or to New Salem Academy, as they had never supported a high school.

The state audited the accounts of the towns by mid-June. No major discrepancies were found, but one former town official of Dana had to hand over some public property he had taken with him. A public auction was held in September to dispose of the public property of the towns, including buildings. This was one of the last of many auctions held in the valley in the past dozen years. Ellis Thayer, a Prescott resident removed to West Brookfield, ran this auction as he had many others in the valley. The old Enfield fire engine brought $67.50. Lots of school textbooks brought a dollar or two per lot. Contractors and wreckers bid on town buildings. The brick Enfield town hall (where the auction was held) brought $550, the grange building only $35. The Dana town hall went for $90; an adjoining schoolhouse got $110. While this auction featured few items that could be called "antiques," in the collectibles sense, many of the previous ones did, attracting collectors and dealers alike. [247]

While the town governments ceased to function, other official activities besides the M.D.W.S.C. ones continued. Most notable of these were the post offices in the valley villages. Both Prescott offices had closed in the 1920's; Greenwich Plains followed suit in 1930, Smith's Village in 1936. Five offices still operated when the towns officially ended; Enfield, Greenwich Village, Dana, North Dana and Millington. All but the first closed in the summer of 1938, as the amount of business shrank to an unprofitable level, or their postmasters moved out. Donald Howe, H. Morgan Ryther, and other philatelists produced numerous "last day of mailing" covers with the appropriate cancel obtained the day each office closed. The Enfield office stayed open longer due to its proximity to the dam and other worksites.

By September, there were few people still living within the valley. All of the renters, and all but a few of the natives had left; a farewell party was held by the Greenwich Lake Campers. Wreckers were busy tearing down the remaining buildings, as sales to outside contractors and rebuilders had ended. Only about a half dozen families had not settled with the Commission on their property. The 41 year old Enfield telephone exchange closed in October. [248]

The few remaining residents and the laborers in the valley had to withstand a hurricane and attendant flood which hit on September 21st. The Swift River again failed to damage the dam, even with more water than in 1936. Since the timber on much of the future flow area had been removed, a lesser degree of damage was suffered there than in adjoining sections. Over three million board feet of mature timber was blown down in the Ware River watershed area; twenty million in the Swift valley. Many roads were temporarily impassable, and work on the dam ceased for a few days because of the storm. The M.D.W.S.C. arranged for the quick cutting and storage of the downed timber with federal assistance, as did many other governmental bodies in New England. [249]

1939 was to be the final year of any human habitation in the Swift River valley. The year began with the closing of the Enfield post office on January 14th. 2,800 last day covers were mailed by postmaster Edwin C. Howe; many to his brother Donald. The postmaster had stayed in the half-torn down building nights in a cold room to keep the office going. The building was converted into a commissary for project workers. [250].

The winter that the post office closed saw the valley turn into a land filled with fire. The clearing contractors were bulldozing trees, brush, and house remnants into large piles, and setting them afire. This was done to speed removal of anything standing within the flow area of the reservoir. As a result, rumors persisting to this day about buildings standing under water are untrue, as all objects more than a few feet high were taken or knocked down.

By May, only three families reportedly remained within the valley. The two Frost families waited at their homes near the West Branch of the Swift for the settlement checks to come in from the Commission. Another family lived in a shack near Greenwich Plains. All around them burning debris were smoking. A series of forest fires led to closing off many of the roads leading into the valley, with guards at the gates. Only those with passes from the Commission could enter. [251]

In July, the dam was completed to the point that it was deemed ready to hold back water when the Commission decided to do so. The same month, a crew from the state division of fisheries and game removed several tons of fish from the ponds in the valley. It was felt that the stocked fish should be relocated where they would be caught by licenced fishermen, instead of living in the reservoir untouched for years. [252]

Despite old stories handed down by some area residents, all people living within the flow area or Commission watershed land were evacuated before flooding commenced on August 14, 1939. No one was removed by boat with water on their front step, or similar follies. A water supply reservoir cannot fill with people living in it s midst creating wastes every day.

While no bands played, no hymns were sung, no speeches were delivered, the day the waters of the Swift began to back up into it s valley was when the four towns of Dana, Enfield, Greenwich and Prescott ceased to exist.

Chapter Six   The River Rises

A couple of weeks after the reservoir began to fill, heavy rains fell for ten days, making the first showing of flooded area impressive. The early rise of the water was eagerly noted by the newspapers, who chronicled the innundation of the Enfield-Belchertown Road bridge site, and the rescue of two wood chucks from a small island soon to be innundated. 253

Crowds came to the administration building at Winsor Dam to view the scenery. They also marveled at the access road being built to connect the dam with the top of Great Quabbin Mountain and the dike. Only a few men were still working on brush clearing in the upper reaches of the valley, but the public did not see them, as access to that area was still limited.

Besides the big contractor's court cases, there were still legal questions involving valley real estate. Hearings were being held by the Board of Referees in late 1939. Hampshire County was compensated by the state $15,000 for costs incurred in suits arising out of the reservoir project. A few individuals received settlements on their property as late as 1945. 254

A curious difference of opinions of the reservoir came out in a 1940 suit brought by the Boston Duck Co. of Bondsville against the Commission. Frank E. Winsor's diary was introduced as evidence, with a quote that the reservoir "improved the flow of the Swift River". Elwood Bean of Rhode Island, an engineer, testified for the complainant that the water coming out of the reservoir would "smell like rotten eggs" in ten years. Unfortunately for Mr. Bean, time has proven him wrong. 255

Early in 1940, a year after the closing of the Enfield Post Office, the last building in that former village was torn down. This was the old Chandler Mansion, which had served as an office and laboratory for the Commission.

While the main concerns of the M.D.W.S.C. had been the completion of the the Swift-Ware water supply system, there were other projects it was assigned to oversee. In 1939, the General Court assigned the Commission the job of building new sewerage systems for the towns of Rutland and Holden, to keep the sewerage out of the Wachusett Reservoir watershed. In 1938, a new aqueduct was authorized to carry water from the terminus of the Wachusett Aqueduct in the Sudbury watershed at Southborough to the distribution reservoir. This aqueduct (later named for M.D.W.S.C. chairman Hultman) was deemed sufficiently complete by October 1940 to hold a formal dedication ceremony.

On October 23rd, Governor Leverett Saltonstall headed a delegation of dignitaries who toured the Metropolitan water works, starting at Quabbin and heading eastward. The Governor started the flow of water through the new pressure aqueduct at Southborough, then went to a formal ceremony at the other end of the tunnel in Weston. There, addresses were delivered by Chairman Hultman, the Governor, Mayor Maurice Tobin of Boston, and P.W.A. Administrator John Carmody. The latter speaker represented over seven million dollars in Federal P.W.A. money granted to the aqueduct project. [256]

A number of other projects to improve aspects of the Metropolitan system would be undertaken by the Commission, including the so-called City Tunnel, and sewerage work on the Metropolitan Boston area. The war years of 1941-45 would see the Commission do no major work, only touching up and maintenance work on the Swift-Ware supplies and aqueducts.

Due to the war footing the nation was placed on during 1940, the public was restricted from access to the Metropolitan reservoirs and guards posted in August, 1940. This was to avert what Chief Engineer Kennison called the "threat of sabotage" from unknown agents of potential enemies. This was later extended in late 1941 to the posting of an M.D.C. police unit at Quabbin Reservoir. [257]

1941 saw many pieces of legislation passed affecting the water situation. An investigation was initiated to determine if other municipalities not in the water district could join it. The use of the Metropolitan police at Quabbin was authorized. Towns receiving welfare cases from the former four valley towns were authorized reimbursement from the M.D.W.S.C. Finally, the Commission itself was enlarged by two members with respect to the Metropolitan sewerage work. The Director of the M.D.C. Sewerage Division and the Chief Sanitary Engineer of the State Health Department were the new ex-officio members. Commissioner Lavelle was replaced by Charles H. Brown upon the expiration of the former's term.

Early in 1941, the tower on top of Great Quabbin Mountain was completed. This was used as a fire spotting tower and radio base, later to be opened for sightseeing use. Also at this time, a petition circulated in the reservoir area to have a commemorative stamp issued in memory of the valley towns. Like many other petitions of this type, it was rejected as being too local in nature to consider for a stamp issue.

In March, officials at Westover Air Force Base in Chicopee proposed using parts of the reservoir and watershed for bombing practice. They felt that Quabbin was the largest area in the northeast available for this type of activity. The M.D.W.S.C. granted the Army Air Corps permission to use the Prescott Peninsula and West Branch flow area. The Air Corps fenced off the northern approaches to the Prescott Peninsula to keep people away from the target area. They also stationed fire fighting equipment and crash boats at the reservoir in connection with this activity. The Army was permitted to use parts of the reservoir basin and watershed land in Rutland for gunnery practice. [258]

In September, 1941, the water level in Quabbin Reservoir was judged sufficient enough to send some of it eastward through the aqueduct. This was done to augment the declining Wachusett supply during a low rainfall period.[259]

World War II saw the Commission reduce its reforestation efforts in the reservoir watershed due to the labor shortage. Sheep were used to keep grass down at the dam for awhile. The use of motor vehicles was cut to comply with gas and rubber tire conservation efforts. Many engineers and other employees were drafted; a half dozen perished in the service of their country.

In 1944 two flood flows on the Ware River were diverted into Quabbin Reservoir, saving downstream property from possible damage. One of these floods was caused by a hurricane, but it did not cause serious damage. In October, the last of 5,761 bodies were transferred out of valley cemeteries into Quabbin Park Cemetery. 260

Eugene Hultman died on April 22, 1945, having witnessed the completion of most of the work that commenced under his Chairmanship of the M.D.W.S.C. He was replaced by William T. Morrisey.

As a result of earlier investigations, a fixed rate of $40 per million gallons was set for water costs for those towns and cities entering the Metropolitan Water District. Because the area of eligibility for municipalities to enter the district had been expanded, Western Massachusetts cities, especially Chicopee and Springfield, were investigating the possibilities of tying into the Quabbin supply.

A contract was made with the New England Power Co. to purchase electricity that would be generated by a hydro-power plant to be installed at the base of Winsor Dam. The work of constructing the plant began in 1945. The electricity was to be used at the administration building, with the surplus to be sold to the power company. Although some had suspected the whole Quabbin project was meant to generate vast amounts of electricity, this was all of the power generation to come out of it. 261

In July, 1945, the area around the dam and dike was opened up to the public, provided no pictures were taken or sketches made. Many came to view the reservoir from the Enfield Lookout between the dam and dike, or from the observation tower on Great Quabbin Mountain. When the war ended a few weeks later, the picture prohibition was lifted.

1946 saw some changes in the M.D.W.S.C. Commissioner Brown died on February 4, and was succeeded by Louis B. Connors. Edward Kelley was replaced by John R. Shaughnessey in September.

Despite the large amounts of water sent to Wachusett, Quabbin Reservoir was considered filled on June 22, 1946. At 2 p.m. that day, the spillway at Winsor Dam was opened to allow the first water to flow through it to the Swift River. Much was made of the flow of water 150 feet down the spillway, and of the fact that the reservoir had essentially reached its 412,000,000,000 gallon capacity. 262

A few weeks later, much of the northern shore of the reservoir was opened to fishing. Fishermen had to walk in from the gates at the access roads, but the trip was considered worth it for the trophy fish that were caught from the start. Agitation would soon begin for boat fishing on the reservoir, to take further advantage of the waters. 263

The reservoir was hardly filled when legislation was passed authorizing the

City of Chicopee and the towns of Wilbraham and South Hadley to take water from it. An aqueduct was constructed for this purpose, connecting Quabbin to the Nash Hill Reservoir supply of Chicopee.

At the beginning of 1947, a hydropower station was under construction at the Oakdale outlet of the Quabbin Aqueduct. An arrangement was made with the New England Power Co. to purchase the excess power generated by this station, as had been done for the power at Winsor Dam.

Most all of the work envisioned for the M.D.W.S.C. had been completed by this time, except the City Tunnel project, which was in progress. Seeing this, the General Court passed Chapter 583 of the Acts of 1947, abolishing the Commission after two decades of existence. The duties and functions of the M.D.W.S.C. were transferred to the Metropolitan District Commission water and construction divisions, the latter being created for this transferal.

The Metropolitan District Water Supply Commission had spent over $70,000,000 of state, local, and federal money in carrying out it's various projects. Over two dozen lives had been lost while the Swift and Ware works were under construction. One hundred square miles of the Commonwealth had been altered by the Commission's work. Goodnough, Smith and Winsor were dead, but their work was there for all to see.

PART V   CUP BEARER TO ALL MANKIND

Chapter One   The Effects of Quabbin

The presence of a filled Quabbin Reservoir in the western part of Central Massachusetts signified many changes beyond the reservoir itself. Such a large body of water gives its region a focal point. This is evidenced by the number of institutions and businesses that adopted the name "Quabbin" after the mid - 1940's.

Local and regional transportation patterns were changed by the reservoir acting as a barrier. The loss of the railroad between Athol and Bondsville caused more use of the highways running north-south along the perimeters of the reservoir. East-west highway traffic was not as affected by the reservoir, as the major routes in that direction had previously been on either end of the Swift River valley.

Land values in the area were also affected by the reservoir. There is evidence that real estate increased in value in the Swift River valley from the end of the First World War until the M.D.W.S.C. started purchasing land at the end of 1926. This follows the anticipation of high purchase prices by the commission for the land it required. A slight slump in values occurred in early 1924, when the General Court appeared unable to find a solution to the water problem. When the M.D.W.S.C. did not rush to buy out the valley in 1927, another decline in land values occurred. They experienced another slight rise in 1928 - 29, only to fall with the arrival of the Great Depression. Those who sold out in the early 1930's got the least for their land. [264]

After the completion of the reservoir, the situation was mixed in the real estate market. Many outlying parts of towns bordering the reservoir watershed were worth more due to the view or forestry attractions provided. Some areas left more isolated by the reservoir cutting off transportation connections lost real estate value. This is especially true of the West Ware and Bondsville areas south of Winsor Dam. In the case of Petersham, this aspect enhanced the summer home orientation of the town. [265]

Another varying effect of the reservoir was it's regulation of the flow of the Ware and Swift Rivers. The M.D.W.S.C. had to pay out over 3/4 million dollars in damages to various waters users along those streams and the Chicopee River. The diversion of some of these waters lowered the amount available for power generation and pollution cleansing. Most of the affected areas have adjusted to this change in water levels, except Bondsville. In this village, the loss of a through rail connection plus lower power generation capacity combined to doom local industry. [266]

There has been a benefit to the control of water by the reservoir, that being flood control. As early as 1944, hurricane induced flood flows in the upper Ware River were diverted into Quabbin Reservoir, thus saving downstream landowners from much damage. The regulations requiring minimum flows in the rivers at all times serves to stabilize flows during dry periods, especially in the lower Swift.

Other ways that Quabbin Reservoir has affected the region less visibly involve population shifts, forestry practices, recreation, scientific research and agriculture.

About 2,500 people were forced to leave the Swift River valley in the second two decades of the twentieth century. Most of them moved to within twenty five miles of the old homestead. The location of relatives, similar jobs, or just a longing to be near their old homes influenced most of these people. Many of the farmers bought farms in towns like the Brookfields, or New Braintree. When the Swift River Box Co. moved from North Dana to Athol, some of the employees moved with it. Some valley people had worked in Athol or Ware, so it was easier to locate where their job was. A few, like Cyprian Uracius of Greenwich, moved their house with them to their new town, rebuilding it there. Many towns around the reservoir gained in population due to the influx of Swift valley people. 267

The existence of so many former valley residents in the vicinity of the reservoir has helped sustain the Swift River Valley Historical Society. The society's museum - headquarters in North New Salem serves as a meeting place and clearinghouse of information for valley refugees, who maintain the property.

Quabbin Reservoir and the Ware River watersheds together comprise over 100,000 acres of land controlled by the M.D.C. and the Commonwealth of Massachusetts. Both watersheds have been reforested by the M.D.C. in attempting to maintain the areas and gain higher water yields. These areas are also managed for timber production, which gains income for the M.D.C. Although both watersheds were overplanted in pine species (as were many such areas in the 1930's), current practices are achieving a more balanced forest. This serves to improve the watershed's appearance, water yield, and wildlife potential. 268

Wildlife has definitely benefited from the presence of the reservoir. Stocking programs of fish have succeeded in the largest freshwater body in the state, making it one of the most popular fishing spots in New England. Experiments with land wildlife have had mixed success. A flock of grazing sheep proved uneconomical and was sold in 1946. That same year, wild turkeys were introduced to the Quabbin watershed, but adverse environmental conditions wiped them out within four years. Later stockings have taken a tenuous hold in the Quabbin watershed. 269

Deer, wildcats, and eagles have found the Quabbin watershed an ideal refuge from the onslaught of civilization. The latter two species might not have survived in Central Massachusetts without this refuge. Geese make good use of the reservoir as a part of their flyway in migration. Because the watershed is larger than any nearby public land holding, it serves as a breeding ground and feeding area for wildlife in adjoining areas.

Quabbin Reservoir has fostered some recreational uses, although it s major use as a water supply reservoir naturally limits this somewhat. The area between the dam and dike has been opened to automobile traffic over service roads. Both dam and dike can be driven over, with picnic areas nearby. There is a lookout spot halfway between the two overlooking the site of Enfield. The firetower atop Great Quabbin Mountain affords a fine view of the whole reservoir.

Although the area known as Prescott Peninsula has been closed to unauthorized access, most of the Quabbin watershed is open for hiking and shore fishing access. All of the old roads leading into the M.D.C. reservation have gates across them barring motor vehicles, but they afford easy routes for hikers.

Shore fishing was opened up on a limited basis in 1946. After much study and debate, boat fishing was allowed beginning in 1952. Three launching ramps are maintained by the M.D.C.; one each in Pelham, New Salem, and Petersham (with access from Hardwick). Strict horsepower limitations on boat motors (eased somewhat in 1981), limitations on the number of boats launched per day, and defined boundaries of fishing areas have minimized the pollution from boats affecting the water supply. It is ironic to note that X.H. Goodnough once wrote an article claiming motor boat use on water supply reservoirs would be dangerous to the water quality. 270 The M.D.C. itself has three boats, appropriately named Greenwich, Enfield and Dana, which have been used at various times for maintenance and law enforcement purposes on the reservoir.

Many people view the limitations on recreational uses of Quabbin as a detriment to public enjoyment. The fact that many more activities are allowed on the Ware River watershed is cited by some in claiming that Quabbin should be "opened up". The M.D.C. has kept a conservative stance on expanding activities at the reservoir, viewing these to be incompatible with the water supply function. Attempts in 1960 and 1976 to establish more recreation through master plans failed after initial public enthusiasm. In prohibiting motor vehicles access, interior camping areas, and hunting, the M.D.C. is probably showing wisdom in managing the area around the reservoir properly. However, in discouraging bicycling, and forbidding cross country skiing, any sailing, and not attempting to set up an excursion boat, the M.D.C. is probably being too tight-fisted about the property. Timber operators are somehow not seen as disruptive to wildlife (although their operations are certainly needed), but cross-country skiers are!

Scientific research has taken place at Quabbin Reservoir since it's inception. Rainfall statistics for the area have long been recorded in connection with the water supply function. Many studies of the aquatic life have been made in connection with fish stocking. The whole watershed area, especially the Prescott Peninsula, has been a living laboratory for official and unofficial flora and fauna research. Students from the nearby five colleges in the Amherst-Northampton area and the Harvard School of Forestry in Petersham have used the Quabbin watershed. Some excellent wildlife photography has been taken there also.

Another kind of research which is not strictly compatible with others done at Quabbin is radio-astromomy. In the early 1970's a facility to receive and analyze such signals was constructed at the site of Prescott Center. The site was viewed as necessary at the time because of it s remoteness from sources of interference to the radio waves. While the facility has done worthwhile research for the consortium of colleges who run it, the poorly supervised access to it (for students) probably disrupts wildlife over much of the peninsula sanctuary.

A final topic in noting the effects of the reservoir on it's surroundings is agriculture. The obvious aspect of this is the removal of over 100,000 acres of land from regular or potential agricultural uses by the land takings in the Ware and Swift valleys. Although less than ten percent of this area was utilized for agricultural uses, [271] it s loss still assisted in the general decline of Massachusetts agriculture. Some of the farmers resettled on similar property in other towns, but there is no evidence they reclaimed a large number of abandoned farms.

What is more important to note in this connection is what effect the reservoir had on areas adjoining it. The most important aspects of this are transportation and markets. The only area wide produce item that may have been affected by the reservoir was berry crops. As Walter Clark noted, blueberry and strawberry production in the Swift valley had been an important sideline for some farmers. [272] With many of the berry patches destroyed or sold because of the reservoir, the collective ability of the producers to ship their product to the cities became crippled. The few remaining berry producers could only cater to a more limited local trade, thus making the fruits less profitable. Apple orchards in the area did not suffer from this problem, as most of these were self-contained operations.

Agriculture in areas on all but the south side of the reservoir was (and still is) rather limited. It did not suffer from the effects of the reservoir. Belchertown, bordering the southwest side of Quabbin, probably had the most developed farming in the area. Because it had a great deal of arable land, good soils, and easy access to Springfield and Northampton, it continued to flourish agriculturally during this period. [273]

The area that suffered most from the reservoir's creation was due south of Winsor Dam, in the West Ware and Bondsville stretch along the Swift River. It has already been noted that cutting off much of the rail and road access out of the area hurt industry there. These same factors adversely affected agriculture in a section with a limited local market with poor soils to begin with. The decline of industry reduced the market and population locally, and poor transportation facilities made it costlier to ship produce to a better market. Bondsville and West Ware have never recovered from the blow the reservoir dealt them. [274]

In assessing the effects of the reservoir, we cannot forget the human factor. How many lives were shattered by the forced removal and loss of familiar environment cannot be determined. There were some who never went back after the project was done, feeling they could not stand the sight of what had become of their valley. Others have gone back and broken into tears upon seeing their old home as a cellarhole overgrown with rows of pine. The natives of the valley will pass away with time, but the bitterness of many will live on to stand as a lesson.

Chapter Two   More Users, More Diversions?

When Quabbin Reservoir was first filled, the Metropolitan Water District consisted of twenty towns and cities. In twenty years, that number would double as more Metropolitan towns would abandon their old supplies and tie in with the M.D.C. Municipalities in the central part of the state tied into the system as well. Chicopee, South Hadley, and Wilbraham benefit from the aqueduct running south out of Quabbin. Six towns near Wachusett Reservoir are supplied by it, and both Worcester and Springfield can use the system in an emergency.

The safe yield of the Quabbin - Wachusett system is currently 300,000,000 gallons per day. This was not meant to serve so many places to the extent required. This results in depletion of the sources eventually if not checked. During the 1950's, while water consumption rose, the amount of rainfall feeding the reservoir was steadily high. After 1960, there was a five year period of declining rainfall, causing the reservoir water level to fall to less than half of it's storage capacity. [275]

The M.D.C. filed a bill in the General Court in 1964 to allow it to study the possibility of diverting water from Millers River into Quabbin to stem the shrinkage. Chapter 606 of the Acts and Resolves approved the study request, and authorized up to $500,000 to fund it. The original deadline of December, 1965 was extended for a year to allow other sources besides the Millers to be studied.

The report issued as Senate Document No. 1095 at the end of 1966 recommended several possible avenues for augmenting the Quabbin supply. The Millers River was seen as being too polluted to utilize, but tributaries such as the Tully River were clean enough to divert. The main focus of the diversion efforts became the Connecticut River.

Even though Connecticut River water was seen as being a class below that of Quabbin there were two strong reasons for the M.D.C. to look to it as a source. One was it s size; it could stand a diversion much better that a smaller stream such as the Millers. The other was a proposal by Northeast Utilities Service Co. to build a pumped storage power project along the Connecticut at the Erving-Northfield boundary.

This pumped storage project was a scheme whereby water from the Connecticut would be pumped up to a reservoir on the top of Northfield Mountain during low power use periods, such as nightime. When power was most needed during the day, water from the reservoir would go back down through turbines in the mountain, thus generating electricity. The M.D.C. reasoned that the reservoir could be made large enough to accomodate a day or two supply of water to divert to Quabbin.

The M.D.C. was given the proposal by the utility company, who saw a way to lessen the operating costs of the plant. The proposal was reviewed by the Army Corps of Engineers, and various state and federal water agencies, including the private Connecticut River Watershed Council. Consensus of the group was that the proposed diversion would not have adverse effects on other interests involved with the river. 276

A resolve was passed by the General Court in 1966 allowing the M.D.C. to continue its studies and to draw up final plans for diversions to Quabbin. Chapter 669 of the same year authorized the M.D.C. to go ahead on the Northfield Mountain diversion aqueduct, allowing up to $25,000,000 for the cost of the project.

Instead of proceeding with the aqueduct construction, the M. D. C. spent $14,000,000 of its appropriation on fixing up parts of its distribution system. A request for more funds in 1969 was met by the appointment of a legislative commission to study the matter. The commission reported back to the General Court in 1970, resulting in another appropriation of $25,000,000 on top of the remaining $11,000,000. The only setback the M.D.C. received in getting more money was that the minimum flow level of the Connecticut under which diversions could commence was raised to 17,000 cubic feet per second from 15,000. 277

At this point, a variety of circumstances intervened to halt the diversion plans from going forward. The most important of these initially was the failure of the M.D.C. and Northeast Utilities to come to an agreement over how the costs of altering the pumped storage project were to be paid. The utility felt that the M.D.C. was reluctant to pay for all the changes to its plant at Northfield. The M.D.C. was not authorized by the General Court to enter into a binding arbitration agreement to settle the question. This gave time for various opponents of the project to build up support for their efforts. 278

Many of the opponents to the diversion got their start during the legislative commission hearings in early 1970. The Springfield Conservation Commission is often seen as an early opponent of the diversion. They in turn influenced the Connecticut River Watershed Council, the Connecticut River Information Clearinghouse, other elements of Springfield government, local Leagues of Women Voters, and the Lower Pioneer Valley Regional Planning Commission. 279

Students and faculty of some of the valley colleges (especially Amherst College and the University of Mass. at Amherst) opposed the diversion and expressed such opinions at the public hearings. However, many of these people came out as being either over emotional or vague in their opinions. These people, joined by others including Dr. Arthur Tamplin of the Lawrence Radiology Laboratory, brought up the issue of the diversion bringing radioactive wastes into the Metropolitan water supply. 280

A few miles upstream from the proposed diversion site a nuclear power generating station is located at Vernon, Vermont. Low level radiation discharges from the plant end up in the Connecticut River. Tamplin, who opposed the Atomic Energy Commission standards for such discharges, felt that 400 to 600 people a year would die from drinking Connecticut River water containing Vernon waste if it was in the M.D.C. system. [281]

Although this issue excited people for a few months, it died out rather rapidly. The same over emotionality and manipulated evidence that failed to convince the legislative commission earlier occurred again. Tamplin and others seemed more concerned with A.E.C. standards on radioactive waste than in their relation to the diversion issue. Many other scientists were in disagreement with Tamplin and his supporters, and the public was hard put to figure out the issue because of its technical nature. [282]

With the failure of the M.D.C. to obtain the right to enter into a binding arbitration agreement with the utility, it looked elsewhere for immediate augmentation of its water supply. The most feasible solution seemed to be the tributaries of the Millers River. The U.S. Army Corps of Engineers maintained a flood control reservoir on the East Branch of the Tully River in Royalston. Water from this, augmented by the equally clean Millers tributaries Tarbell and Priest Brooks, would have added 48 million gallons of water daily to Quabbin.

The Army Corps enthusiastically took up the plan, along with the Northfield proposals, in its Northeast Water Supply Study (NEWS). Hearings on the study held in the Millers River basin did not show the local populace to be enthusiastic about the plans, but the study was pursued. Both proposals ended up as recommendations in the study report. Federal funding could not be obtained for the projects, however, causing the Army Corps to withdraw its endorsement of them in 1979. A 1980 Federal water projects funding bill that would have included the Northfield project had an amendment tacked onto it eliminating the scheme from the funding. This amendment had been the work of Congressman Toby Moffit of Connecticut.

While the Army Corps support for its projects made the M.D.C. hopeful of getting them built, there were still problems on the state level as the seventies dragged on. A decision to run extra water through the outlet at Quabbin Reservoir for power generation was criticized. The M.D.C. stated that its early 1974 action was in response to the national energy crisis, but the Connecticut River Ecology Action Corp. charged that it was a way to keep Quabbin water levels low for justifying a diversion into it. [283]

The Dukakis administration (1975-79) and its Environmental Affairs Secretary Evelyn Murphy, were not sympathetic to the diversion proposals. They required the M.D.C. to participate in a statewide water resources study and other impact studies before proceeding any further with diversion plans.

One study got the M.D.C. embroiled in controversy in 1975. This was the so-called Curran Report, as it was subcontracted to the Curran Associates firm of Northampton, Mass. by the University of Mass. - Amherst. This study found that up to 48 million gallons a day of water was being leaked out of the distribution system in Boston alone. This was calculated as being almost half of what the whole system was leaking. Although the study has been criticized for the extent of its estimates, it gave ammunition to diversion opponents, and has caused the subject of repairing leaky pipes to be considered in later studies.

Ironically, at the time of the release of the Curran Report, Quabbin Reservoir had reached its highest levels in fifteen years. Nearby Amherst and Belchertown were not as water rich. Both towns had been experiencing rapid growth due to the expansion of the campus of the University of Mass. in the 1960's. In seeking additional water supplies, both towns noted that Quabbin would be an alternative. M.D.C. officials did not seem reluctant to offer water to the towns, as it would add two to the four municipalities along the Connecticut River receiving M.D.C. water. The more towns along the river tied to the M.D.C., the less opposition to diversion from that area. The towns ended up rejecting the M.D.C. offers. 284

As a reaction to the continuing diversion proposals, two steering committees were set up late in 1976. In October, a Connecticut group formed a committee to oversee uses of the Connecticut River. This group was bolstered by a resolution in that state's legislature opposing the Northfield project. In late November, the Massachusetts Steering Committee for the Connecticut River Valley was formed with representatives from all over the river valley. The latter group was definitely opposed to the diversion plans, and called for the valley to be the master of its own resources. 285

Early in 1977, the draft of a state wide water supply study was published. Since the co-ordinator for the study was the chief planner for the M.D.C., it was not surprising to many Western Massachusetts observers that the Northfield project was seen as "most cost effective, least environmentally disruptive, and most certain to achieve the desired augmentation results". The report was also criticized for not calling for citizen input on water allocation, and for empasizing the needs of the M.D.C. communities. 286

The final report, issued a year later, recommended that the M.D.C. undertake full impact studies on all possible alternatives to their perceived water problem. The diversion of water from the Western part of the state was viewed as a "last resort solution" after conservation and other efforts failed. The report had the approval of Secretary Murphy and Governor Dukakis.

At the time of the report's issuance, the General Court passed a resolution endorsing the recommendations of the report regarding any M.D.C. diversion plans. A few weeks later, the Connecticut River subcommittee of the New England River Basins Commission altered its earlier support of the diversion to place it at a last resort status. The full Commission adopted this policy at the end of 1978. 287

Another slap at diversion plans was delivered at the end of 1978 by Lt.

Governor Thomas P. O'Neill, III. He presented figures to the Federal government indicating that the M.D.C. needed $400,000,000 to repair leaky pipes in its distribution system. He felt that the sum would be better spent in repairs than on diverting water from the Connecticut River. No Federal money was forthcoming from the request, but it did underline the scope of the leakage problem. [288]

Late in 1979, the M.D.C. negotiated a contract with a consulting firm for the two year environmental impact study of the viable alternatives to solving their water problems. In January, 1980, the firm of Wallace, Floyd, Ellenzweig, and Moore was retained to do the study for $740,000. The Northfield Citizens Advisory Committee, created in 1977 by Evelyn Murphy, charged that the M.D.C. was trying to keep it from monitoring the study as closely as had been promised. The committee was given access to participate in the study, although not as extensively as desired.

The first phase report of the consultants study was released in October, 1980. The desalinization of sea water alternative was rejected for a number of reasons, but most of the other alternatives were retained for further study. These included diversion of the Millers, Connecticut, or Merrimac Rivers, conservation, Quabbin watershed management, and development or reclamation of ground and surface water in M.D.C. communities.

As the second phase of the impact study continued into 1981, the M.D.C. responded to pressure that it stress conservation of its water resources. The M.D.C. voted to discourage municipalities from applying to join the water district, and to endorse conservation attempts by communities. Various regulations were beefed up or relaxed to encourage conservation. [289]

If through the impact report recommendations a diversion of water into Quabbin is achieved, then the story of the reservoir will continue as an ongoing part of Boston's water supply history. If water is obtained through conservation or pipe repairs, then the people of Central and Western Massachusetts can rest a bit easier in the hopes that water politics will not intrude on them again for a time.

PART VI   APPENDICES

Part 1   Footnotes

For newspaper abbreviations, see Bibliographical Essay. In a few cases, one note may cover more than one paragraph.

1. Pub. Doc. 147, 1935 (M.D.W.S.C. Annual Report).   2. Kennison article in 6/34 NEWAJ pp 217-8.   3. Ibid.   4. Underwood, Quabbin... p.31.   5. J.G. Holland History of Western Mass. Vol.II (1855) p.202..   6. Marion Starkey A Little Rebellion (1955) pp. 64-5.   7. Holland, op. cit. p.204.   8. L.P. Coolidge Past Events of Prescott (1949) p.7.   9. Gay's Gazeteer of Hampshire Co. (1887) p. 278, Holland, op. cit. p. 214.   10. Holland, op. cit. p.202.   11. Dana Centennial, p.30-1.   12. Holland, op.cit p.269.   13. Underwood, op.cit. p.195.   14. Ibid, p.220.   15. Town tax assessment lists.   16. W. Lord History of Athol, (1953) p. 194.   17. 1925 Mass. census.   18. G.A. Wilson, unpub. paper Death of the Swift River Valley (1972) p.16.   19. 1885 Mass. Census.   20. Underwood op.cit. p.267.

21. W. Clark Quabbin Reservoir (1946) p.304.   22. L. Klimm, population paper, (1933) pp. 121,124,66.   23. Clark, op.cit. p.32.   24. Ibid., p.33.   25. Ibid, pp. 33-7.   26. Federal Censuses, 1850, 1890,1920.   27. D. Howe, Quabbin: The Lost Valley (1951) pp.383, 443.   28. Interviews, Schmidt 1976, Hansen, 1981.   29. Clark, op.cit. pp.182 -3; Willy interviews, 1976.   30. Springfield Sunday Republican, 8/8/65.   31. E. Gustafson, Ghost Towns 'Neath Quabbin Reservoir (1940) p.14. 32. N. Blake Water for the Cities(1956) p.5.   33. Statutes, Vol. II 19 -21.   34. F. McInnes articles in NEWAJ 3/32, p.8.   35. N. Blake. op.cit. p.194.   36. Ibid, p.195. 37. Ibid., p.197.   38. Ibid, pp.208 -9.   39. Ibid, p.215.   40. C. Bowen Yankee From Olympus (1943), p.88.

41. A General Description of the Water Supply of the Boston Met. Dist. (1940), pp.7 -9. 42. 1895 Mass Census, 1880 Federal Census.   43. H. No. 1550, 1922, p.13.   44. N. Blake, op.cit. p.268.   45. H. No.1550, 1922, p.   .   46. 50th Anniv. Sketch of Harvard Class of 1882.   47. L. Haskins, Abundant Action (1968), p.31.   48. Goodnough card in the file of Boston Soc. of Civil Eng. 49. Haskins, op.cit., p.31.   50. H. No.500, pp. xxxv., 121.   51. Ibid., p.xvIII.   52. Quoted (from H no.500?) in Howe, op.cit., p.23.   53. H. No.500, 1895, p.131.   54. WT, 9/1/94.   55. H.M. Hamilton History of West Boylston (1956), pp.125,140.   56. H. Wasserman Harvey Wasserman's History of the U.S. (1972), p.217-8.   57. C. Weidner Water for a City (1974), p.252.   58. Wasserman, op.cit., p.52-3.   59. R. Wiebe, The Search for Order (1967), pp.172, 181. 60. Howe, op.cit., p.vii.

61. AT 12/99.   62. AT 7/20/09.   63. H No. 1550. 1922, p.98.   64. Ibid, p.112 -4. 65. Ibid, p.52.   66. BP, Jan., 1920.   67. H No. 1500, 1922, p.4.   68. AT 2/21. 69. AT 2/22/21.   70. AT 3/22/21.   71. AT 3/8/21.   72. AT 5/10/21.   73. AT Ibid.   74. H No.1550, 1922, pp.5-10.   75. Ibid, pp.29-37.   76. Ibid, pp.177-8, 204, 239.   77. 1922 Manual of the General Court.   78. E.Adlow, Threshold of Justice, (1973), p.86.   79. BGL. 9/19/76.   80. R. Sawyer, Two Decades in Ware (1929), pp.7,14.

81. AT 5/29/24.   82. L. Haskins, op.cit., p.29.   83. Ibid, pp.29 -30.   84. Haverill

Gaz. 5/24/23.   85. Ibid.   86. AT 2/21, 3/21/22.   87. BGL, 5/19/22.   88. SR 3/25/22.   89. AT, 4/18/22.   90. AT 5/9/22.   91. SR, AT 5/19/22.   92. SR 5/23/22.   93. AT 5/30/22.   94. SR 5/28/22.   95. SR 6/7/22.   96. Ibid.   97. NEWA Journal, 6/22.   98. AT 7/11/22.   99. AT 7/29/22.   100. SR 3/23/23.

101. SR 4/11/23.   102. H No. 1724, 1924,p.15.   103. AT 7/31/23.   104. AT 1/15/24, SU 1965 quoted in G.A. Wilson, op.cit.   105. SR 1/29/24.   106. AT 4/1/24.   107. AT 7/29/24   108. SR 5/11/24, AT 5/13/24.   109. H No.1724, 1924, p.9.   110. Ibid, p.15.   111. Ibid, pp.15-6.   112. AT 5/20/24.   113. SR 5/20/24,   114.   114. SR 5/28/24.   115. AT 6/3/24.   116. H. No.900, 1926, pp.76-7.   117. AT 11/17, 12/18/25.   118. SR 1/18/26.   119. SR 3/3/26.   120. AT 1/19, 2/9/26.

121. SR 2/18/26.   122. SR 2/19/26.   123. Ibid.   124. SR 2/24/26.   125. WT 2/24/26.   126. WT 2/26/26.   127. WT 3/3/26.   128. SR 3/3/26.   129. Ibid.   130. SR 3/3/26.   131. WT 3/6/26.   132. WT 3/9/26.   133. SR 3/17/26.   134. WT 4/2/26.   135. WT 4/3/26.   136. WT 4/7/26.   137. WT & SR 4/8/26.   138. Ibid.   139. Alan Fox interview.   140. SR & WT 4/9/26.
141. SR 4/9/26.   142. WT 4/9/26.   143. SR & WT 4/10/26.   144. SR 4/10/26.   145. WT 4/14/26.   146. SR 4/17/26.   147. AT 4/20/26.   148. WT 4/27/26.   149. SR & WT 4/28/26.   150. Ibid.   151. WT & SR 4/29/26.   152. WT 5/1/26.   153. SR 4/17, 5/1/26.   154. SR 5/4/26.   155. WT 5/4/26.   156. SR. 5/6/26.   157. WT 5/7/26, SR 5/8/26.   158. SR & WT 5/11/26.   159. WT 5/4/26.   160. SR 5/15/26.

161. SR 5/17/26.   162. SR 5/19/26.   163. SR 5/9/26.   164. SR 5/21/26.   165. WT 5/21/26.   166. SR & WT 5/26/26.   167. L. Haskins, op.cit, p 34.   168. W. Clark, op. cit, p.185.   169. AT 8/3, 11/9/26.   170. AT / / .   171. AT 12/26/26.   172. AT 1/12/27.   173. AT 1/18/27.   174. AT 2/1/27.   175. AT 2/8/26.   176. C.M. Saville brief for State of Conn., 1927.   177. AT 2/15/27.   178. SR 3/17/27.   179. AT 3/23/27.   180. SR 3/23/27.

181. SR 4/6/27.   182. "BACKSITE" No. 4.   183. Christenson Article in NEWA Journal 6/40, p.207.   184. Winsor in Kennison Article in Ibid, 6/34.   185. WT 8/24/34.   186. WT 5/4/28.   187. AT 12/6/27.   188. AT 1/24/28.   189. AT 7/10/28.   190. AT / /30.   191. Winsor Article in NEWA Journal, 10/31.   192. Quotes from testimony on 282 US 660.   193. Opinion 282 US 660,   194. 283 US 336.   195. AT 2/4, 8/19/31   196. AT 6/20, 7/1, 7/7/31.   197. AT 10/1, 10/13/31.   198. AT 9/16/31.   199. AT 11/5/31.   200. AT 6/15/32.

201. AT 8/10/32, Kirkpatrick int.   202. L.P. Coolidge, op.cit, p.59.   203. AT 5/6/32, SR 10/26/32, Kennison Article in NEWA Journal 6/34.   204. Christenson op.cit, 6/40, p.207, NEWA Journal /45, p.141.   205. Ibid, /.44.   206. Ibid.   207. AT 3/8/33.   208. AT 11/3, 9/13/33.   209. AT 10/25/33; Christenson, op.cit. 6/40.   210. SU 7/27,28,29, 8/7/33.   211. P.D. 147, 1934.   212. Ibid, SR 4/28/33.   213. AT 4/28/, 9/19, 11/28, 12/19/34.   214. AT 5/2, 6/6, 11/21; SR 10/28, 12/3/34.   215. J. Dineen, *The Purple Shamrock* (1949) p.232.   216. BGL 4/23/45, SR 1/15/35.   217. Dineen, op.cit. pp.232-4.   218. AT 6/5, 7/31/35, SR 9/25/35.   219. SU 12/29/32.   220. E. Irey *The Tax Dodgers* (1948), p.p. 284-9.

221. ADN 8/25/36.   222. SR 4/29/36, Willey interview.   223. HG 4/12/36, WRN 3/9/38.   224. SR 5/7, 6/5, 7/7/36.   225. SR 10/22, 9/14/36.   226. SR 12/2,

11/24/37. 227. WRN 12/15/37, SR 2/24/37. 228. WT 11/15/37. 229. WT, WRN, 3/9/38 HG & SR 4/9/38. 230. SU 7/28/38, HG 5/10/38. 231. SU 6/27/38. 232. SR 8/13, 9/1, 9/19/38. 233. WRN 9/28/38. 234. SU 12/31/38. 235. Boston Adv. 1/31/39. P.D. 147, 1939. 236. SR 2/7, 6/16, 7/9/39. 237. WRN 5/17/39. 238. BG 9/17/36. 239. E. Gustafson, op.cit., pp.82-5. 240. BG 4/12/36.

241. SR 8/2/36, 150th Ann. program of Enfield Cong. Church. 242. Ibid, p.9, SR p.9, SR 8/10/36. 243. P.D. 147, 1937, WRN 9/18/37 244. WRN 1/12/38. 245. SR 4/22/38. 246. SU 4/28/38. 247. HG 6/16/38 SR 9/14/38, Gustafson, op.cit., pp.102-110. 248. 9/18/38. 249. P.D. 147, 1938. 250. SR 1/15/39. 251. WRN 5/17/39. 252. WRN 7/19/39. 253. SR. 8/26/39, 1/5/40. 254. SR 10/11/39, 2/28/40, Ch. 42, Resolves of 1945. 255. SU 3/8/40. 256. HG 10/24/40: *A General Description*, op.cit. 257 WT 8/15/40. 258. SU 3/28, 8/10/41, P.D. 147, 1941. 259. SU 9/17/41. 260. P.D. 147, 1944.

261. Ibid, 1946. 262. HG 6/23/46. 264. Christenson, op.cit. 1940, pp. 199-200. 265. D. Kellogg, *The Swift River Drainage Basin and Quabbin Reservoir* (1951, unpub. paper) pp.32-5. 266. Ibid, p.7. 267. Mass. Censuses, interviews, S.R.V.H.S. Roster. 268. B. Spencer interview. 269. P.D. 147, 1946, SU 11/27/50. 270. NEWA Journal 3/19. 271. Christenson, op.cit. 1940. 272. Clark, op.cit. pp.34-7. 273. Kellogg, op.cit. pp. 57-8. 274, Ibid. pp.34-5. 275. S No. 1095, 1966, pp.4, 10. 276. Ibid. p.6., J. Spivack unpub. paper *Massachusetts Water Policy* (etc.), 1978, P.42. 277, Ibid, p.53. 278. Interview with NU official (anon.) 1978. 279. Berger (ed.) Pub. No. 28, Water Resources Research Center, U-Mass., Amherst (1971) P.A.-8. 280. Ibid, P.A-12.

281. BGL 5/22/70 282. Berger, Ibid., P.A-14-16. 283. SR 2/19/74. 284. Valley Advocate 12/1, 12/15/76. 285. Ibid, 12/1/76. 286. Valley Advocate & HG 6/9/77. 287. WT 12/25/78. 288. ADN 12/12/75. 289. M.D.C. Water News 6/81.

Part 2   Bibliographic Essay

Books

Since Quabbin Reservoir is considered either local history or an engineering feat, there has been little published in book form on the subject. The three standards are: Walter Clark's Quabbin Reservoir (1946, Ky.) which emphasizes the technical aspects of the construction and landtaking (Clark was an appraiser); Evelina Gustafson's Ghost Towns 'Neath Quabbin Reservoir (Boston, 1940), an oversentimental travelogue of the valley's last days; and Donald Howe's Quabbin: The Lost Valley (Ware, Mass. 1951). The latter book was written by several hands, and contains some minor errors, but it is well illustrated and still serves as the best general work on the subject. The author's own Atlas of the Quabbin Valley: Past and Present (Athol, Mass. 1975) serves as an introduction to the subject or a supplement to the Howe book.

Only Enfield and Prescott have had formal town histories written about them. Francis Underwood's Quabbin: The Story of a New England Town (Boston, 1893) is a social history of Enfield written without naming individuals. Lillie P. Coolidge's Past Events of Prescott (Orange, Mass., 1949) is not much more than a compilation of news clippings and old town records, although it serves the genealogist well. The closest thing to a Dana history is the book printed for the town's centennial celebration Dana Centennial (Barre, Mass., 1901). Dana is the only one of the four towns with a published volume in the "vital statistics" series.

Histories of nearby towns consulted for sidelights on early area history and reservoir connections include Athol, Petersham, Barre, Ware, Pelham, and New Salem. Unfortunately little has been written about the villages of the Ware River watershed.

Few biographies have been written about the leading figures in this story. Joseph Dineen's book on James Michael Curley, The Purple Shamrock (Boston, 1949) was helpful for background on state politics in the 1930's. Both Leslie Haskins and Roland Sawyer had short autobiographies privately published; the former's Abundant Action (1968) and the latter's Two Decades in Ware (1929).

Most histories of Boston are not very useful for this subject. Three books stand out in their coverage of the Boston water story. Brackett's history of 1868, and Desmond Fitzgeralds history covering 1868-1876 are good for the early years. Nelson Blake's Water For The Cities (Syracuse, N.Y., 1956) is a good comparative study of the development of northeastern U.S. water supplies up to the 1920's.

Essays and Articles

Periodical literature on the Quabbin Reservoir is numerous, but varies in quality. Magazines like Yankee run an article on the subject every few years or so; these are best when they include recollections of valley natives. Articles covering everything from fishing on the reservoir to the methods of landtaking surveys were consulted. Those found useful are cited in the footnotes. The New England Waterworks Association Journal (NEWAJ) provided many useful articles on historical and technical aspects of the story, as did other engineering journals.

Many newspapers have run feature articles on the subject since the reservoir was completed. Most of these are written by people with little understanding of the subject beyond the basic facts (Material on the diversion plans in the 1970's is an exception). Some of these articles do contain useful recollections of valley natives.

Government Publications

Publications of the Commonwealth of Massachusetts and its various branches were invaluable in compiling material for this work. The Acts and Resolves of the General Court (formerly called Private and Special Statutes of the Commonwealth) contain all of the legislation passed in relation to the subject. The Legislative Documents for relevant years contain many of the bills and reports filed concerning water issues; the 1895 book contains the "Stearns Report" advocating what became the Wachusett Reservoir, while the 1922 book has the Joint Board report (H. No. 1550) complete. The daily records of the House and Senate are useful in following the course of a bill to passage or defeat. The Manual of the General Court and Public Officials of Massachusetts contain information on state legislators and their districts at the time of issuance.

Most of the commissions and agencies involved with water supply are required to issue annual reports to the state which are published. Reports used here include those of the Boston Water Board (1846-95), Metropolitan Water (and Sewerage) Board (1895-1919), Massachusetts Board (later Dept.) of Health (1886 -), Metropolitan District Comm. (1919 -) and the Metropolitan District Water Supply Comm. (1926-47).

A mass of documents has appeared concerning the diversion issue. Of note are those issued by the Army Corps of Engineers during it's involvement with the issue in the 1970's, and M.D.C. studies for the impact statements in progress as this book was being written. The Massachusetts Water Supply Policy Study (Boston, 1977) by an M.D.C. employee, gives that agency's view as well as any document.

The M.D.C. has privately issued studies on forestry and master plans for Quabbin, among other subjects. A 1972 study (Baron, et al.) by the Construction Division advocated rebuilding the rail line from New Salem to Athol to haul salable fill out of the reservoir bottom! Such studies are of interest mainly to the specialist in the relevant area of study.

An excellent source of material on the Supreme Court case between Connecticut and Massachusetts is the collection of testimony and briefs filed in connection with the case. The M.D.C. library has the only set of the multi-volume series seen by the author.

Newspapers

A list of newspapers consulted in compiling this work appears below. Rough time periods are included to indicate what part of a paper's file was useful. The most consistent coverage of the subject through this century has been in the Springfield newspapers (the Union and Republican). The Athol, Barre, and Ware papers often gave good local coverage of the reservoir story. Geographically close, but with very little coverage of the topic, is the Amherst Record. Abbreviations used here correspond with the footnotes.

Athol Chronicle AC (1920-35), Athol Transcript AT (1895-35), Athol Daily News ADN (1935 to date), Barre Gazette BG (1930's), Boston Globe BGL (1920's, 1938), Boston Herald BH (1920's), Boston Post BP (1920's, 1938), Greenfield Recorder GR (1920's, 1970 to date), Daily Hampshire Gazette DHG (1936 to date), Orange Enterprise and Journal OEJ (1920's), Springfield Daily News and Morning Union SU (1920 to date), Springfield Republican SR (1920-40), Ware River News WRN (1920-1940), Worcester Telegram WT and Evening Gazette EG (1924 to date).

Interviews

The author wishes to thank all those who consented to be interviewed. Some of these people provided information directly incorporated into the text, while others confirmed information or helped provide a good background on various aspects of the story.

Former or present employees of the M.D.W.S.C. / M.D.C. interviewed were Mr. Gerard Albertine, 1978; Mr. Stanley Dore, 1978; Mr. Karl Kennison, 1976; Mr. Roger Lonergan, 1980; Mr. Bruce Spencer, 1976, and Mr. Harold "Red" Willey, 1976.

Former residents of the Swift River Valley and surrounding region interviewed were: Mrs. Bessie Currier Brown, 1980; Mr. Lester Hager, 1977; Mr. Herman Hanson, 1981: Mr. and Mrs. Frank Kirkpatrick, 1976; Mrs. Annis LaPlante, 1980; Mrs. Eleanor Schmidt, 1976; Mr. Cyprian Uracius, 1980; and Mrs. Marion White, 1978.

Some of these people are deceased; the time for us to hear the voices of those who were there is passing quickly.

Final Note

There are three places that stand out as depositories of information on Quabbin Reservoir and the towns it affected. The M.D.C. Library in the M.D.C. headquarters building in Boston has many hard to find documents relating to engineering studies and legal matters. Captain Albert Swanson has assembled a good collection.

The M.D.C. administration building at Winsor Dam is the depository for town records of the four inundated towns. There is also a collection of photographs of all properties purchased, along with photographs of many construction projects. Numerous maps and plans of properties and projects are also here, as is the Quabbin Visitors Center.

The Swift River Valley Historical Society has a wealth of material on the four towns and New Salem. Its museum (open Wednesday afternoons, 2-4 p.m. July and August) is in the charming old Whittaker-Clary House on Elm St. in North New Salem. Artifacts, photographs, town records, and genealogy materials are available for study or perusal.

The author is considering conducting further research on this subject, and is especially interested in obtaining original valley photographs or post card views. Write in care of the author, J. R. Greene, 33 Bearsden Road, Athol, Mass. 01331

Part 3 Population Tables

YEAR	BOSTON	DANA	ENFIELD	GREENWICH	PRESCOTT	NEW SALEM
1765	15,520			434		375
1790	18,320	(Inc. 1801)	(Inc. 1816)	1,045	(Inc. 1822)	1,543
1800	24,937			1,460		1,949
1810	33,787	625		1,225		2,107
1820	43,298	664	873	778		2,146
1830	61,392	623	1,056	813	758	1,889
1840	93,383	691	976	824	780	1,305
1850	136,881	842	1,036	838	737	1,253
1860	177,840	876	1,025	699	611	957
1870	250,526	758	1,023	665	541	832
1880	362,839	736	1,043	633	460	869
1890	448,477	700	952	526	376	845
1900	560,892	790	1,036	491	380	807
1910	670,585	736	874	452	320	630
1920	748,060	599	790	399	236	512
1930	781,188	595	497	238	48	414
1935	817,713	387	495	219	18	443
1940	770,816	0*	0*	0*	0*	357

*Four towns legally abolished in 1938. Greenwich and New Salem lost population in the early 1800's due to losing territory to other towns. Boston gained territory and population by annexing several towns in the late 1800's. All but the 1765 and 1935 figures are from Federal Censuses: the others are State figures.

Part 4   Index of Persons

Adams, Samuel, 2
Adlow, Elijah, 25, 56, 57
Allen, Frank R., 25
Allardice, E., 34
Alling, Benjamin, 64
Andrus, Rev. J.C.   80
Arnold, H.B.   41
Averill, Ernest   64

Bacon, Gaspar   42, 49, 50
Bailey, James A., Jr.   16, 20, 23, 28, 29, 30, 31, 32, 34, 36, 37, 39
Baldwin, Loami   9
Ballou, Hosea   4
Barlow, Ellis & George   67
Bean, Elwood   85
Booth, George   38, 40, 42, 45, 46. 47, 49
Brooks, W.H.   37, 42
Brown, Charles H.   86, 87
Bunn, Charles W.   64, 65
Butler, S. Ct. Justice   65

Callahan, G.J.   64
Carmody, John   85
Clark, Walter   58, 92
Connors, Louis B.   87
Cook, George W.   58
Cooke, Charles   54
Coakley, Daniel   75
Coolbeth, Herbert   82
Coolidge, Calvin   31, 68
Cox, Gov. Channing   35, 38
Curley, James Michael   71, 72, 73, 75, 76
Cutler, R.A.   64

Dana, Judge Francis   3
Davis, Joseph P.   12, 13
Davis, Rep.   44
Davenport, Charles   52
Day, John P.   67
Dearborn, Attorney   37
Doane, Fred   81

Duggan, John   68
Duncan, William   54, 55
Dukakis, Michael   96, 97

Eliot, Mayor   10
Ely, Gov. Joseph   71
Evans, Wilmer   15

Farley, F.D.   53
Felton, Charles   34, 54
Field, Robert   3
Foss, W.E.   43, 44
Freeman, John R.   15, 18, 42, 46, 47, 49, 51
Fuller, Gov. Alvan   43, 44, 45, 51, 57
Frost, Families   84

Galvin, William   67
Gilman, Rep.   43
Gilmore, Charles   41
Glazier, Rep.   33
Goodnough, X.H.   12, 13, 18, 19, 20, 21, 22, 23, 26, 27, 28, 30, 31, 32, 33, 34, 36, 37, 39, 43, 44, 45, 46, 47, 48, 51, 52, 53, 64, 73, 88, 91
Gow, Charles   38, 40, 41, 42, 44, 45, 46, 48
Gow, Frederick   38
Greenwich, Duke of   1
Griswold, Senator   30, 32
Gustafson, Evelina   7

Hager, Arthur & Otis   73
Haigis, Senator   35, 49, 50
Hamburger, Rep. Leo   30, 31
Hammond, N. LeRoy   53
Hapgood, Ben   37
Hartshorn, Senator   47
Haskins, Leslie T.   26, 27, 41, 46, 47, 51, 53, 54
Hazen, Allen   38, 39, 40, 42, 43, 46, 48, 64
Hinds, Captain   2

Hobson, E.E. 29, 55
Holden, Bendict 64
Holmes, Justice Oliver W. 65, 66
Howe, Donald 83, 84
Howe, Edwin C. 81, 84
Hultman, Eugene 71, 72, 75, 77, 85, 87
Hurley, Gov. Charles 77, 82

Innes, Charles 48

Johnson, J.H. 28, 30, 51, 54
Johnston, Lewis 81
Johnston, Lillian 81

Kelley, Edward 75, 87
Kelley, Dr. Eugene 20, 21, 29, 31
Kennison, Karl 20, 43, 44, 53, 64, 79, 86
Kenniston, Davis 40, 42, 43, 44, 48, 49, 50, 51, 52, 55, 56, 57, 60, 71
Kimball, N.B. 55, 56

Lavalle, Thomas 74, 86
Leahy, Francis 78
Lochridge, Elbert E. 38, 42

McCormack, John 25
McLane, Sen. Walter 35, 47, 49, 50
Main, Charles T. 20, 52
Matthews, Nelson 51
Maher, Thomas 68
Merriam, Thaddeus 77
Merrill, Randolph 80
Mellish, Attorney 51
Moffitt, Cong. Toby 96
Molt, R. Nelson 52
Moriarty, James 78
Morrissey, William 87
Muir, John 16
Mullen, Rep. 41
Murphy, Evelyn 96, 97, 98

Nani - (or Nine) Quaben 1
Nelson, Sen. Christian 33, 34, 35, 41, 43, 49
Nichols, Mayor 42, 46

O'Day, John T. 75
O'Neill, Thomas P., III 98

Parker, Herbert 40, 41, 42, 47, 48, 49
Parsons, F.E. 51
Pike, Rep. Chester 35, 36, 51
Pond, George 29, 56, 60
Putnam, Judge F. Delano 78

Quincy, Josiah 9

Rice, Senator A.B. 42, 47, 49
Ryther, H. Morgan 61, 83
Roosevelt, Theodore 16, 26

Saltonstall, Leverett 25, 85
Sanford, Samuel 61
Saville, Caleb M. 44
Sawyer, Rev. Roland D. 26, 28, 30, 32, 33, 34, 35, 36, 37, 38, 40, 42, 46, 47, 50, 51, 53, 54, 56, 57, 60, 79
Schoonmaker, John D. 17
Schoonmaker, John H. 29, 37, 41, 48, 55
Segur, Dr. Willard, 54, 81, 82
Schaughnessy, John R. 87
Shays, Captain Daniel 2, 72
Shattuck, Henry 25, 37, 38, 51
Smith, Almond 26
Smith, J. Waldo 20, 23, 31, 34, 44, 48, 52, 64, 65, 88
Smith, John D. 73
Soliday, Joseph W. 52
Stearns, Frederic 12, 13, 15, 18, 20
Storrs, George 41, 48, 54, 55
Swan, Rep. Clyde 67

Tamplin, Dr. Arthur 95, 96
Thayer, Ellis 83
Tobin, Maurice 77, 85
Treadwell, Daniel 9
Trumbull, John 43, 44
Twohig, Rep. 51

Underwood, Francis 5
Walcott, Dr. Henry P. 12, 13, 16, 20, 34
Walker, Charles 81
Walker, W.H. 58
Walker, Stephen 82
Walsh, Senator David 61
Warner, Joseph 64
Warren, Bentley 64
Webster, George P. 21, 27, 29, 30, 31, 32
Wells, Senator Wellington 33, 35, 36
Weston, Arthur 53
Weston, Robert 48
Wheelwright, J.W. 29
Whipple, Professor 31
Winsor, Frank E. 34, 52, 53, 56, 71, 85, 88
Worrell, Thomas 54, 55

Young, B. Loring 40